Fishing boat designs: 4

Small steel fishing boats

FAO
FISHERIES
TECHNICAL
PAPER

239

Prepared by
David J. Eyres[*]
Fisheries Technology Service
Fishery Industries Division

FOOD
AND
AGRICULTURE
ORGANIZATION
OF THE
UNITED NATIONS
Rome, 1984

PREPARATION OF THIS PAPER

The paper contains designs of two general purpose steel fishing boats and general guidance on boatbuilding in steel plus cost estimating for fisheries officers, vessel owners and boatbuilders having experience in working with other materials.

Acknowledgement is made to G. Breekveldt, Marine Architects Ltd., P.O. Box 2642, Auckland, New Zealand, for the designs shown and Adriana Barcali and Maurizio Carlesi for draughting and layouts.

* Present address: c/o Marine Division
Ministry of Transport
Private Bag
Wellington, No. 2

ABSTRACT

This, the fourth of the FAO fishing boat design handbooks, deals with the subject of small steel fishing boats.

The publication is intended to serve individuals and companies with experience of structural steel fabrication, boatbuilders working with other materials who may wish to build larger craft in steel, and finally fisheries officers and vessel owners who require guidance on boatbuilding in this material.

Section 2 describes the material, the building site and tools required, fabrication and outfitting of steel boats.

Section 3 contains information on protection against corrosion and Section 4 presents a simple method of cost estimating to construct a small steel craft.

The handbook also contains designs of 15 m and 21 m general purpose steel fishing boats. These consist of general arrangements, lines and hull construction drawings, steel quantities and outline specifications.

Distribution:

FAO Fisheries Department
FAO Regional Fishery Officers
FAO Fisheries Field Projects
Selected Boatbuilders and Naval Architects

For bibliographic purposes this document should be cited as follows:

Eyres, D.J., Fishing boat designs: 4. Small
1984   steel fishing boats. FAO Fish.
Tech.Pap., (239):33 p.

FISHING BOAT DESIGNS: 4

SMALL STEEL FISHING BOATS

Contents

1. INTRODUCTION

2. BUILDING IN STEEL . 1

   2.1 The Material 1
   2.2 Premises and Site 2
   2.3 Tools and Equipment 2
   2.4 Marking Out or Lofting 4
   2.5 Fabrication and Erection 5
   2.6 Outfitting 7

3. PROTECTION AGAINST CORROSION 8

   3.1 General 8
   3.2 Cleaning and Priming Steelwork 9
   3.3 Paints and Application of Paint Systems 9
   3.4 Bi-metallic Corrosion 10

4. COST ESTIMATING

ANNEX
   Outline Specification for 15m and 21m Steel Fishing Boats 16

LIST OF FIGURES

1	Bending Press	19
2	Lofting Details	19
3	Sacrificial Anodes	20
4	Net Steel Weight	21
5	Manhours (Steelwork)	21

LIST OF TABLES

		Page
1	Weld Details	22
2	Typical Paint Systems	25
3	Steel Quantities for 15m Vessel - Mild Steel Plates	26
4	Steel Quantities for 15m Vessel - Mild Steel Sections	27
5	Steel Quantities for 21m Vessel - Mild Steel Plates	29
6	Steel Quantities for 21m Vessel - Mild Steel Sections	31

LIST OF DRAWINGS

For 15m Steel Fishing Boat
SB1- 1	General Arrangement
SB1- 2	Lines
SB1- 3	Offsets, Framing and Plating Diagram
SB1- 4	Hull Construction I
SB1- 5	Hull Construction II
SB1- 6	Sections
SB1- 7	Deckhouse
SB1- 8	Details of Keel, Skeg and Sternpost
SB1- 9	Sterntube and Shaft Details
SB1-10	Rudder, Stock and Tube Details
SB1-11	Details of Sliding Door

For 21m Steel Fishing Boat
SB2- 1	General Arrangement I
SB2- 2	General Arrangement II
SB2- 3	Lines
SB2- 4	Offsets, Framing and Plating Diagram
SB2- 5	Hull Construction I
SB2- 6	Hull Construction II
SB2- 7	Deckhouse
SB2- 8	Sterntube and Shaft Details
SB2- 9	Rudder, Stock and Tube Details
SB2-10	Details of Watertight Door

For 15m and 21m Steel Fishing Boats
SB1/2-1	Typical Details of Deckhouse
SB1/2-2	Typical Details of Bulwark and Berthing
SB2/2-3	Mechanical Steering Gear
SB1/2-4	Fish Hatch
SB1/2-5	Small Hatch Details
SB2/2-6	Typical Details of Longitudinal Framing System

1. INTRODUCTION

This publication in the FAO Fishing Boat Design series deals with the construction of small steel fishing boats. By small steel boats is meant fishing vessels of less than say 30 metres overall length where the required building facilities and equipment are modest and the construction techniques somewhat different from those of larger shipbuilding practice. Generally the construction of boats in steel is not considered below say 12 metres in length where steel is at a weight disadvantage and below 15 metres in the case of tropical marine conditions because of corrosion rates on the thinner steel plate used. The design of two different general purpose steel fishing boats of 15 m and 21 m overall length are presented.

Simple hull forms are utilized to avoid the use of sophisticated equipment and skills and the text provides information on the material and its maintenance, the necessary building equipment and the important principles which differentiate steel boatbuilding from general steel fabrication practices. It is not the intention to teach the basic skills of welding and gas cutting, a knowledge of which is assumed and is not uncommon in developing countries. The purpose of the publication is rather to show how these skills can be adapted to steel boatbuilding. Detailed information on construction techniques is not within the scope of the booklet and the notes are necessarily brief, but an attempt has been made to cover the more important points and the drawings contain a great deal of extra detail to help the inexperienced builder.

For steel hulls of less than 30 m in length, the equipment and building facilities required are modest and can be compared with the traditional wood boatyard rather than the highly automated steel shipyards producing larger vessels.

Notes on estimating steel weights and costs of steel fishing boats are also included. Finally the layouts of the two designs presented are discussed in detail.

The two steel fishing vessel designs may be suitable for various fishing operations within the Exclusive Economic Zones (EEZs) of selected developing countries but should be adapted to local fishing conditions. For practical reasons full working drawings and details of construction could not be included in this paper. The designs comply with the FAO/ILO/IMO Guidelines for the Design, Construction and Equipment of Small Fishing Vessels, published by IMO. Scantlings are in accordance with good practice for this size of craft and should meet the requirements of most regulating authorities. The Organization, however, takes no responsibility that this will be the case and the onus is on the builder to meet his legal responsibilities for plan approval and survey where this is necessary.

It is also recommended that safety precautions specific to steel fabrication, erection and outfitting be followed during the construction of the boats. Guidelines from the ILO publication "Safety and Health in Shipbuilding and Ship Repairing" should be followed.

2. BUILDING IN STEEL

2.1 The material

Steel as a building material for boats is strong and easy to work. Welded joints are equal in strength to the basic material if welders have sufficient skills and some practice in making watertight structures. It is also easy to repair by cutting out and welding in new material using facilities available almost anywhere. Steel does not burn and will last for a long time providing proper protection against rusting is maintained.

Small craft, mainly these 15 m length overall, with thin steel plates, are very sensitive to corrosion especially in tropical conditions. They need constant and time consuming maintenance by regular rotational cleaning and repainting of the parts of the hull subject to rust attack. Precaution against corrosion on the immersed parts of the hull require the boat to be put ashore frequently for cleaning and maintenance. The less accessible areas of the steel structure are troublesome to protect and require care both during the design and construction phases to avoid the formation of inaccessible rust breeding pockets. High maintenance costs of the hull in tropical environments should be taken into consideration before a decision on construction of a steel fishing boat is made.

A wide range of steels are marketed in the form of plates and a variety of sections. For boatbuilding, plain low carbon steels which are reasonably priced and readily available in most countries are more than adequate. It is recommended that the boatbuilder orders from the steel supplier hot rolled mild steel plate and sections complying with British Standard BS 4360, Japanese Standard JIS G3101 or US Standard ASTM A131-74 or equivalent standards.

If a boat is designed to the rules of a particular regulating authority or classification society, this should be clearly stated when ordering material to meet the chemical and physical properties required by the rules. Such rules should then be observed during the design, construction and trials of the boat.

Sections required for the two designs in this publication are limited to flat bars and equal or unequal angles with some hollow rolled sections. There is no reason for the boatbuilder to use a wide variety of sections.

A list of steel materials for the 15 m and 21 m designs in the handbook is given in Tables 3, 4, 5 and 6. The quantities listed are net quantities and the percentages to be added to give the quantity to be ordered are indicated below the tables. When ordering plates for the hull and deck it is good small steel boatbuilding practice to order the largest size of plate that can be handled to reduce waste and cut down the amount of working and welding required. The wastage figures given with the tables are based on this assumption and if for some reason you are not able to utilize the largest plates, add another 5 percent for wastage. Study the body line body plan and deck plans and decide how the largest available plates may be utilized before ordering. Also plan the cutting of floors, brackets, etc., from standard plates to avoid excessive waste.

It is advisable to inspect the steel on arrival for defects particularly the plates for flatness and delaminations. Reject any material that is not satisfactory and could give trouble during fabrication and erection. Steel plates and profiles should be stored properly to avoid bending under their own weight and material should be kept in sorted packages for easy access.

2.2 Premises and site

Ideally the hull should be built within covered premises protected from the weather. However, this is not always possible and temporary protection may need to be provided for the critical welding operations during unfavourable climatic conditions.

The vessel should be erected on ground which can support its weight and which may need reinforcing for this purpose. It is preferable that the erection site be adjacent to water unless very heavy lift and transport facilities are available.

If a number of vessels are to be built as a new enterprise the installation of a permanent launching way may be justified. This can be done for end launching by levelling a gradient of about 1 in 10 reinforcing the surface and arranging launching rails to below the low water mark. The expense of fabricating a permanent steel cradle to support the vessel during construction and for subsequent launching may also be justified. For a 'one-off' construction, unless a heavy lift is available, temporary but adequate means of support and launching should be considered before commencing building.

2.3 Tools and equipment

The tools and equipment required for a small steel boatyard are not extensive and need not cost any more than those required for a comparable wooden boatyard. If general steel fabrication is already undertaken, then the additional equipment is minimal.

2.3.1 Steel cutting equipment

For this purpose one or two oxygen/acetylene hand torch sets are sufficient for the size of boats covered by this publication. Some items of this equipment should be specially selected for boatbuilding use. The torch should be a combination unit so that tips for other purposes than cutting can be used. A light torch with tip at right angles to the handle is the most suitable. A two stage regulator is preferable to a single stage regulator. There should be sufficient length of hose to reach any part of the boat during construction with the gas bottles remaining on the ground. A light hose is preferred for this purpose and should be that manufactured for this purpose and supplied with the cutting equipment. A hand cart for transporting the gas cylinders will be found useful.

2.3.2 Welding equipment

The first consideration relates to the availability of an electrical supply at the site and if available the nature of that supply. When no electricity is available, where the current is inadequate or where appreciable voltage fluctuations occur, a petrol or diesel motor driven welding generator unit must be used and this could be utilized for other power tools and lighting.

If a satisfactory mains electrical supply is available a wide range of commercial welding machines of the preferred type can be obtained to operate on 230/460 volt three phase 50/60 cycle supplies. The machines may be of the A/C motor D/C generator or quieter rectifier type both of which incorporate cooling fans. For use in tropical climates it is important to ensure that this cooling is adequate to protect the machine. It is recommended that the boatbuilder should only consider using a conventional direct current (DC) machine with manual flux coated stick electrodes for building boats in the size range covered by this handbook. For this purpose a machine capable of delivering up to 200 amps should be adequate but versatility in the selection of current is desirable to permit welding of a range of plate thicknesses. Multiple weld runs will be used for the heavy plate thicknesses.

If a motor driven generator unit is to be used it should be noted that ordinary electric generators are not suitable for arc welding. A properly designed arc welding generator must be used if it is intended to make up a welding set using an existing petrol or diesel motor, and the motor has to be matched to the generator for correct power and speed. If at all possible a complete made up unit by a reputable manufacturer should be purchased.

The number of units or output terminals required will depend on the number of welders employed at one time. On the 21 m boat at least 3 welders could be employed on final welding up if the boat is to be completed in a reasonable time.

Electrode and earth return (ground) cable sizes are related to current carried and length and should not be of greater length than necessary which would make them difficult to handle and inefficient. Use short lengths with cable connectors so that unneeded lengths may be taken out of the circuit.

The electrodes used should produce a deposited metal which is as close in composition as possible to the parent metal. For the conventional mild steel construction envisaged in this handbook a general purpose electrode capable of depositing weld metal in all positions should be used. A reputable supplier of electrodes can give advice on the availability and types of such electrodes. Store the electrodes in a dry space in sealed containers and if the atmosphere is very humid and wet or any dampness is suspected before use, then bake the electrodes in an ordinary oven at say 150ºC.

When buying welding equipment for the first time the boatbuilder should examine carefully all the available equipment, seek advice from local established steel fabricators if possible and deal with a reputable manufacturer of commercial welding equipment.

2.3.3 Lifting and plate handling equipment

Handling steel plates and the larger sections will necessitate the use of some mechanical hoisting and hauling devices. Chain falls preferably of worn geared type are recommended for lifting and a couple of ratchet action level hoists are very convenient for pulling plates up into position or closing gaps. An old motor truck which can be used around the building site if not on the road may be fitted with steam pipe or similar sheer legs and a hand winch on the tray. This has proved to be a good investment and versatile for lifting jobs around many yards. Clamps of various types are available to provide lifting points on plates and the safest of these is that having a positive locking grip. Crow bars are useful for lifting plate edges and solid steel round or pipe rollers can be helpful in moving plates and heavy sections.

2.3.4 Press

A small hydraulic press which can easily be made up by the boatbuilder is shown in Fig. 1. This can be used to put light curvature in the forward frames and beams as necessary.

2.3.5 Miscellaneous tools

A range of small tools will be found necessary in fabricating and erecting the vessel. These will include iron G cramps (smaller than those required for wood boatbuilding), heavy duty electric power tools with grinding and sanding discs if possible, chipping hammers for veeing thicker steel plate edges before welding and removing weld slag, 3 kg or similar hammers for fabrication work plus steel measuring tape, spirit level and plumb bobs.

2.4 Marking out or lofting

The offsets (i.e. reference dimensions for the hull shape) for the two hull designs incorporated in this publication have been faired. The body plans (Lines) in Drawings Nos. SB1-2 and SB2-3 give the offsets and each transverse frame section can be drawn out full size on a suitable surface. The line so obtained corresponds with the toe of the angle frame (i.e. the inside of the plating) and the frame can be made accurately to this line. It is recommended that unlike a conventional lines plan where only half the section is drawn the full transverse frame section should be drawn, i.e. port and starboard sides should both be drawn which will give greater accuracy and make it simpler to fashion floors as well as the frames, beams and brackets.

The full size frame lines can be marked off on any suitable surface which will not distort and on which the lines can easily be sighted and picked up when necessary. Sheets of plywood painted white or even steel plates tacked together are commonly used. If steel plates are used the main reference points should be punched up to give them some permanence. As the majority of the frame lines are straight they can be drawn in with a straight edge between the reference points or struck in by a chalk line. Forward of the deck break where there is curvature in three or four frames, the reference points should be plotted and a fair line drawn through them with a wood batten pinned or weighted to the marking surface. It is important that the full size framing plan be drawn accurately and the vertical centre line should initially be constructed perpendicular to the base line by striking off arcs with a large beam compass made up with a wooden batten, nail for centering and marking pencil or similar arrangement (see Fig. 2). Offsets are to be accurately set off from the vertical centre line and horizontal base line. The deck line in section is cambered and the standard camber curves are shown in Drawings Nos. SB1-3 and SB2-4. Given the height of deck at sides which may be joined by a straight horizontal line and the height of deck at centre line for any frame the standard camber can be set off for that height above the deck at side line. A fair curve drawn through the camber curve points using a wood batten as for the curved frame lines will give the molded deck line which is the line of toe of the deck beam. Apart from the frame and beam sections the stem curvature has to be lofted out full size and also the profile of the stern assembly so that the correct setting of the sole piece and stern bar and rake of propeller post are obtained.

Straight frame angle can be marked off directly on the marking surface. To obtain the forward frame curvature a piece of 12 mm copper tube can be used as a template and bent to the marked frame line. The angle bar frame can then be bent in the hydraulic press to match the tube template and subsequently checked against the lofted frame line. The 12 mm copper tube may also be used to provide a template for the stem bar curvature and beam curvature if necessary. When checking the curved beams against lofted sections mark on the beams the vessel's centre line which will be found useful when erecting the sections.

Bulkheads, floors and brackets can be drawn in on the full size plan and if available sheets of a suitable thin opaque material can be laid over these to make patterns which can be transferred to the plate. A number of these patterns can be nested onto a plate to give the least wastage when floors, etc., are cut.

2.5 Fabrication and erection

Each frame section with beam should be fabricated and presented to the lofted frame lines to check correctness of shape. Tack weld the components together first and fully weld when checked. Re-check after welding and straighten if necessary.

The transverse frame sections and bulkheads when erected will form the framework or 'built in' jig on which the platework and interior steelwork will subsequently be assembled. For the two boats in this handbook we recommend that the frames be erected in the vertical plane the correct way up. There are a number of builders of steel boats who advocate building the hull in the inverted position which has advantages but has the major disadvantage that it has to be turned up the right way eventually. This would be a major operation with either of these two boats and it is likely to be very difficult for the small builders for whom the book is intended.

The method of fabrication of smaller hulls in turned up position can be relatively easy if rotary jigs are used. These are not very expensive and quite simple in construction and application. Such a set-up facilitates not only the fabrication but avoids much overhead welding which is of great advantage from the point of view of quality and costs.

Unless a large number of similar boats are to be built there is no advantage in building a steel jig on which the plating is formed before the framing is inserted (another practice adopted by some experienced builders).

A number of structural members can be fabricated or assembled prior to erection, for example bulkheads with stiffeners and horizontal stringers or floors with stiffeners and face bar also engine seats with web, bed plate and stiffeners.

When commencing erection the flat bar keel should be set up on blocks with its rake measured above a horizontal line or wire strung along the blocks. The frames and floors can then be measured off along the horizontal wire. Make sure that the keel is set up at a height which will permit you to work underneath the hull comfortably when applying the bottom plating. With the keel in its correct position the stem is set up and plumbed vertical to ensure it is in the same plane as the keel then adequately supported with temporary props. The pre-fabricated aft peak bulkhead can be set up in place and transverse frame and bulkhead sections erected above the keel bar but plumbed perpendicular to the wire or line on which they have been marked off. A wire stretched between the stern and aft peak bulkhead centre line at top is used to check the common centre line of transverse frame sections and bulkheads.

Stern frame fabrication and erection requires special attention because of the heavy sections to be welded. The arrangement should be lofted as mentioned in Section 2.4 and the components sole piece, propeller post, stern bar, boss and stern tube cut out on the loft floor. The lower piece of the propeller post should be cut 6 to 10 mm oversize and then trimmed exactly to length on erection when the boss and sterntube is aligned on the boat. The sole piece may require some heating to put the knuckle in it. To illustrate erection of the stern frame let us take the 21 m vessel as an example (see Drawings Nos. SB2-5 and SB2-6). The erected aft peak bulkhead D and floors C and C/D should have oversize cut-outs for the stern tube and a wire put through in the usual manner to represent the shaft centre line which in this case is parallel to the keel line. The boss and sterntube assembly is aligned to the wire shaft centre line through the cut-outs and has a welded flange on the gland end which can be tack welded to the bulkhead to locate the sterntube at that end. The lower portion of the propeller post is plumbed (with spirit level) on the sole piece which has been tacked in place to the keel and can be trimmed to the correct height for centering the boss. The length of the upper portion of the propeller post is not critical. The lower propeller post is tacked in place with bracket to sole piece and bracing to ensure it is in the same plane as the keel and the boss is then tacked to it. The stern tube may also be lightly connected to the two floors with plate filling pieces. The upper propeller post with stern bar and connecting bracket which may be assembled and checked on the loft floor should then be erected, plumbed and checked for height above the base line at the transom before being propped in place. The transom and other sections between the aft peak bulkhead and transom can then be erected. Welding up the stern frame requires special care to avoid distortion of the shaft line. The sterntube should only be tacked up and not fully welded until as much as possible of the structure and shell plating in this area is fully welded. Both pieces of the propeller post are 'veed' before erection using a chipping hammer to permit full penetration welds at the boss and sole piece. Note that the propeller post is chamfered on the aft face if possible to avoid a blunt trailing edge detrimental to hydrodynamic performance, and, in the case of the 15 m vessel, on the forward face of the propeller post to reduce it to the same thickness as the skeg plate to obtain a satisfactory weld connection (see Drawing No. SB1-8). Get the steel stockholder to do this for you if you cannot handle it. A large number of manual weld runs will be necessary in view of the thickness involved and these should be done with alternate runs on each side of the assembly and alternately top and bottom to minimise distortion.

The initial erection of keel stem and stern assembly and transverse framing with bulkheads should be done very carefully and it is worth taking time to check that everything is accurate at this stage. Do not proceed further until entirely satisfied with the correctness of the erected structure.

The positions of the longitudinals can be marked on the midship frame stern or forward frames as shown on the hull framing diagrams (Drawings Nos. SB1-3 and SB2-4). Each longitudinal frame line can be faired on the erected transverse frames which are then marked and the cut outs made before the longitudinals are fitted and welded in place. With the longitudinals in place you have a very rigid faired framework on which to assemble the plating.

On the 15 m vessel the single skeg plates are templated and fitted at the boat after erection of the propeller post and should be carefully welded in similar sequence to that for other members of the sternframe (see Drawings Nos. SB1-4 and SB1-5). On the 21 m vessel the double plate skeg is double first, that is before any shell plating is fitted. The skeg plates are carried 12 mm or so above the inner surface line of bottom plating (see Drawing No. SB2-5) so that a good weld connection is made and for this the floor should be suitably notched at the intersection of the plates. When plating this skeg you will need to put in a wedge or filling piece in way of the rabbet at the keel (which is 25 mm thick) where it meets the propeller post which is 60 mm thick. The side plates should then be knuckled so they fit to both the keel and propeller post at this point (see Drawing No. SB2-5).

Plating of the hull should be carried out systematically and carefully for accuracy. Hardboard or similar templates lifted off the hull will give best results. If you plate the sides first it is easier to get at the inside to clamp and tack plates in place. The bottom plates can be lifted and shored or chocked in place. Plates should be fitted alternatively to port and starboard to avoid pulling the centreline structure out of shape which may occur if first one side and then the other is plated. In large vessels it is usual to carry out as much of the installation of steelwork and heavy outfit work in the hull as is practicable before starting the deck plating.

For smaller vessels there can, however, be advantages in putting on the deck first provided large access holes are left. Work above the deck can be carried out at the same time as that below and a plated deck provides some shelter from the elements when building outside.

Welding the hull plating should be carried out carefully and in a balanced fashion to avoid distortion as mentioned above. Concentrate on the longitudinal seam welds first starting off at amidships and working towards both ends. Start with the seams at chine and sheer. Maintain the same amount of weld each side of vessel and do not try to do too much at any one time. Keep the weld runs short.

When the shell plating has been welded, particularly the longitudinal welds, the frame to plate welds (which are intermittent welds) are made. Framing should not be continuously welded to the plating which will cause appreciable distortion. The intermittent welds are extremely strong if compared with other methods employed for fastening frames in boatbuilding. Bulkheads should be left until last because continuous welds to the shell are required to maintain watertightness and unsightly distortion can occur if this is not done carefully. Use short lengths of weld about 35 mm at a time with cooling in between welds.

Although more straightforward than the hull plating, the deck plating should also be carefully and systematically fitted and welded before the deck beams are welded to it by intermittent welds.

Welding details for steel boats are given in the welding schedules in Table 1.

2.6 Outfitting

Fitting out the machinery space of a steel boat is probably easier than in boats of other materials, it being relatively simple to cut and weld attachments and seats for items which are compatible with the hull material. The outfitting of the accommodation is more difficult because the linings and furniture are inevitably of a different material. Connections for linings, etc., are normally made to wood grounds which are strategically connected by bolts through the framing or bylings welded to the framing. The number of holes drilled in frames, stiffeners and beams should be kept to the minimum possible and lugs are preferred where the framing member is a critical strength carrier. Details of lining connections should be illustrated in the working drawings. The degree and standard of lining out will depend on the quality of finish required by the owner and any regulations concerning fire resisting materials and accommodation standards. Insulation behind the linings is optional depending on climatic and habitability conditions.

The insulation of fish holds in a steel boat deserves special mention. The 15 m vessel has a dry fish hold with insulation material placed between the frames after coating the steel with a bitumastic paint (see Drawing No. SB1-1). Wood battens are bolted to the frames (or to steel lugs welded to frames) and further insulation material placed between the battens. Linings may be of sheet metal (aluminium or galvanised steel) or plywood with a fibreglass skin. The linings may be screwed to the battens or where a metal lining is used a compatible metal flat bar is sometimes screwed to the wood battens and the lining welded to the bar to give a seam free surface.

The use of chilled sea water tanks in the 21 m vessel complicates the construction somewhat (see Drawings Nos. SB2-1 and SB2-2). For these tanks the drawings indicate steel liners welded to steel lugs which have been intermittently welded to the frames, thus the lining of the tanks is well within the line of the side frames. Other vessels have and are being built with fibreglass liners over insulation material sprayed onto the hull or fibreglassed plywood liners over insulation at the hull sides. Considerable loads are imposed by the tank contents and the author has experience of fibreglass plywood liners giving way after a short period of service hence the preference for the more structurally sound steel liners. Most European builders appear to weld the liners directly to the heel of the frames but for areas where higher sea temperatures are experienced the heat loss with this arrangement is too high. It is preferred that the steel liner be within the heel frame to reduce the heat transfer through the frame in vessels which are to fish in warmer waters. The insulation should be put up in preformed slabs. Foamed in place insulation is prevalent for void spaces. Non-combustible insulation material should be used so that the liners may be welded to the steel lugs after installation of the insulation. Be cautious of fumes from the heated insulation during the welding operation.

Installation of machinery particularly the main engine is best left to engineers experienced in this work and the builder may wish to sub-contract this work. Note only that the vertical main engine bed plates (longitudinal bearers) with heavy flat bars on the top should be integrated with the hull structure and possibly welded to the bulkheads at each end (See Drawing No. SB1-4).

Foundations of all machinery should be strong and stiff enough and be welded to the hull structural members.

The mechanical steering system for the two boats shown in Drawing No. SB1/2-3 incorporates a wormgear speed reducer of standard type which can be readily purchased from most machinery agents. A reduction ratio of between 10 to 1 and 15 to 1 is generally used. The speed reducer should have an extra bearing on the output shaft. The standard speed reducer wormgear has a right hand thread and if installed the wrong way round the ship will turn in the opposite direction to the wheel.

Propeller shafts for steel craft are normally made of bronze or stainless steel with bronze liners working in ferrobestos lubricated bushes (see Drawings Nos. SB1-9 and SB2-8).

They are usually tapered at both ends and keyed to match the propeller and the half coupling.

If the main engine is situated well forward an intermediate shaft of forging quality mild steel should be mounted with half couplings or forged flanges to match the propeller shaft and engine crankshaft flanges.

If a nozzle is being manufactured care should be taken to obtain a smooth shape and the design clearance between the nozzle and propeller tips should be accurately maintained.

The pipe systems on a steel boat do not differ from those used on boats constructed in other materials but it is easier to fix them as holders can be welded to stiffeners. Welding holders directly to the shell, bulkheads, and decks should be avoided as far as possible as this could injure locally the strength of the main hull structure if not properly done. Deck equipment must be mounted in strengthened areas and for heavily loaded machinery and rigging reinforcements under the deck should be allowed for in the hull construction drawings.

Small items of deck equipment can be welded over beams or local reinforcing brackets, angle or flat bars can be added under the deck.

The rules for electric wiring on small steel vessels are the same as for other boats. Cables are fixed to cable trays welded to the structure. The cabling should avoid water contact, not being sited low in the bilge and where possible kept below deck or otherwise given full protection. Cables exposed to the weather should be led in pipes or inside masts and where a penetration through the deck or any other watertight partition is necessary an efficient gland is important.

3. PROTECTION AGAINST CORROSION

3.1 General

One of the greatest drawbacks to the use of steel in building smaller fishing vessels particularly for tropical areas is the materials' tendency to rapidly corrode in sea water if not adequately protected and maintained. Unfortunately maintenance of commercial vessels in developing countries is often minimal and the steel boatbuilder from the start should make the assumption that the boat will not receive as much maintenance as it should once it has left his yard. This infers that a greater margin for corrosion on scantlings may be in order and application of the initial protective systems in the boatyard needs to be of a high standard. Every effort should be made in the construction to avoid areas which are difficult to get at for maintenance and painting purposes. The addition of steel structures for cosmetic purposes which add to the maintenance requirement should be discouraged and the structure limited to the purely functional.

It has been found that the composition of mild steel within the practical range for structural plating has little influence on its corrosion rate in sea water. Thus the boatbuilder should be wary of claims for special and expensive mild steels.

3.2 Cleaning and priming steelwork

Steel plates and sections are generally formed by the hot-rolling process during which the surface becomes oxidised and they leave the steel mill with a thin coating of iron oxides or mill scale. It is very important that this mill scale and any rust, grease, dirt or other surface pollution is removed before applying any surface coating. The ideal means of ensuring complete cleaning of the steelwork surface is to sandblast the hull after erection and immediately apply a zinc or aluminium based priming paint. This priming coat should have the necessary film thickness to give adequate cover of the roughened blast-cleaned surface.

Sandblasting is an unpleasant job and can be hazardous particularly within the hull but is well worthwhile despite the greater expense and should be undertaken if at all possible even if only for the external hull. It is not usually necessary to invest in such equipment as it can be hired in many countries. Particular care should be taken to ensure the equipment is in good condition and that only the correct silica sand is used.

The application of the priming paint should be done as quickly as possible after the surface has been sandblasted. In humid countries evidence of rusting may re-appear after only one hour, therefore a limited area of the hull should be sandblasted followed by immediate priming before proceeding with further areas. The interval between sandblasting and priming should never exceed 2 hours and the paint should be hard dry before further sandblasting is undertaken. It may be possible to buy in steel which is already blasted and primed but generally this can only be done when placing large orders with the steel mill.

If sandblasting equipment is not available then the millscale and surface pollution has to be removed as completely as possible by other means. Plates which have stood in storage exposed to the weather for a lengthy period will be heavily rusted and this rust may be removed by grinding, careful scraping and wire-brushing by hand, which will take much of the millscale with it. It cannot, however, be guaranteed to remove all the mill-scale. If available, flame cleaning is a better alternative to hand cleaning but care should be taken not to overheat the plates, which may change the physical properties of the steel.

Note that if the surface is not sandblasted then advice should be sought about the priming paint subsequently applied since many modern high duty paints are formulated only for application to well prepared surfaces. Paint manufacturers may be reluctant to give advice on this subject for obvious reasons and if faced with this problem a traditional paint system should be applied and not a sophisticated high duty system.

3.3 Paints and application of paint systems

The very important requirement for good preparation and priming has been dealt with in the previous section. Any paint system being only as good as the prepared surface to which it is applied. Paint manufacturers will provide the builder with details of complete paint systems for steel boat hulls which may be divided into three basic categories. Firstly, there are the traditional systems with conventional bitumastic and aluminium or lead paints, secondly the more sophisticated one component (or pack) systems using say chlorinated rubber or vinyl paints and thirdly the sophisticated two component systems using epoxy paints. If a traditional system is used it will almost certainly need renewal after one year's service. The one component system will give two years service and possibly three and the two component system will give three years' service. A traditional system is best suited to poorly prepared steel and a construction yard with limited facilities. The one component system may give limited protection under such circumstances but in this case the two component system should not be considered at all. For an established yard regularly building fishing vessels with sandblasted steel the second system is commonly used mostly based on a chlorinated rubber paint system.

Table 2 indicates typical steel boat coating systems including deck and house as well as the hull. It is debatable as to whether it is necessary to apply anything more than the cheap traditional paint system to areas other than the hull in fishing vessels.

Anti-fouling paints which prevent marine growth on the underwater hull come in different strengths and formulations and if the vessel is to operate in tropical or semi-tropical conditions where fouling could be heavy, ensure that a super tropical or similar anti-fouling paint is used. Longlife anti-fouling paints should be used with the high duty paint systems if the benefit of longer periods between drydockings is to be achieved. Follow the manufacturers recommendations as to the minimum and maximum times allowed before the newly applied anti-fouling coating is immersed.

Internal spaces below deck should also be painted carefully in accordance with paint manufacturers recommendations. Particularly prone to corrosion are the bilges and steel areas behind fish hold linings and inside the skeg which should be treated with bitumastic coatings. The insides of fresh water tanks should be coated with a substance which will not taint the water. A cement wash is often applied. Fuel tanks should be left unpainted internally and covered by a thin layer of oil after cleaning.

Other internal surfaces can be primed with traditional lead or aluminium based paints and finished with gloss enamel paints.

3.4 Bi-metallic corrosion

A simple electro-chemical corrosion cell is formed by two different metals immersed in an electrolyte solution (sea water) and connected externally. The bronze propeller and steel hull of a fishing boat in sea water are a classic example of such a cell. There will be a small conventional electric current flow from the cathode (the propeller) to the anode (the steel hull) at which corrosion will take place. To prevent corrosion in such circumstances it is normal practice to fit sacrificial anodes to the hull in way of the propeller and in way of other immersed bi-metallic structures. Generally such sacrificial anodes are of high purity zinc which is more anodic than steel, i.e. current flow is between the propeller and the zinc anode and preferential corrosion of the zinc anode occurs.

Information on the fitting of zinc anodes should be sought from the manufacturer if possible but the main point is to ensure that the zinc is of as high a purity as possible. A guide to the amount of zinc anodes to fit and location is given in Fig. 3. Welding lugs are normally cast into the zinc block so that they may be attached to the hull. Make sure they are not painted over after fitting.

4. COST ESTIMATING

Before building a steel vessel the parties involved will wish to know with reasonable accuracy the likely cost. From the builder's point of view it is imperative that the cost estimate be accurate if he is to stay in business in a competitive economy. There is no magic formula by which the cost can be immediately found and the more accurate the desired estimate the more detailed will the cost data have to be and a greater breakdown of material and labour costs will be needed. This section looks at the question of cost estimating for steel boats of the type presented in this handbook and indicates desirable data to be collected and applied. Cost data will vary considerably from one country to another and within countries, therefore only local knowledge can provide the builder with the final figures to meet his own situation. For anyone building a steel boat for the first time our advice would be to make the costing as detailed as possible. With more experience and data short cuts in the costing may be adopted.

Firstly, if we consider the basic steel costs it is normal practice to estimate weights and subsequently steel costs based on net weights of steel used in previous vessels. The net weight is converted to an invoiced weight for this purpose by adding a margin for wastage. The net weights of steel for boats having similar scantlings and proportions to the two designs in this handbook are given in Fig. 4. An invoiced weight for steel may be obtained by adding 10 percent wastage to the net weight of steel plate and 5 percent wastage to the net weight of sections. Where the builder is inexperienced it would be prudent to increase these margins to 15 percent and 10 percent respectively. The average local cost per ton of steel plate and sections may be obtained from the supplier.

For greater accuracy in estimating the steel costs the average cost per ton of plate and sections may be considered separately since there is often a significant difference (see Tables 3, 4, 5 and 6). Also a number of very heavy rolled sections may be considerably more expensive than the rest and these can be added separately. These sections may, in fact, be quoted on a cost per length basis rather than per ton.

Having obtained the total material cost for steel, the cost of welding consumables, cutting gases, etc., may be estimated as a percentage of this total. An average figure would be 15 percent.

Labour costs for the steelwork and other work are determined from the recorded hours worked on previous boats. Any efficient boatyard will keep detailed records of the hours worked by different traders on various components of the boat. For comparative purposes when dealing with steelwork the total hours may be converted to a manhours per ton of steelwork ratio. The total hours worked by a small experienced steel boatyard in a developed country constructing boats similar to those in this handbook are shown in Fig. 5. The manhours recorded by inexperienced boat builders in countries where materials and tools, etc., are not readily to hand and climatic conditions are difficult may prove to be 50 to 100 percent higher. For preliminary costings an average figure for manhours/ton for the total hull and house steelwork may be sufficient. For greater accuracy the boatbuilder may sub-divide the steelwork into components where there can be a marked variation in manhours depending on the complexity of the work. Typical subgroupings are:

(1) Framing and Bulkheads
(2) Hull Plating
(3) Deck and Hatches
(4) Deckhouses
(5) Bulwark and Belting
(6) Tanks
(7) Mast, Boom and Gantry.

Manhours/ton on the last two items are much higher than on the other items.

Knowing the manhours required for the steelwork this may be converted to a labour cost by multiplying the total manhours by an hourly wage rate. This average hourly wage rate is not usually that paid to a single tradesman employed on the boat but is in effect what is sometimes referred to as a 'charge out' rate for labour. That is the total wage bill for the boatyard plus cost of ancillary benefits paid to employees and charges to the yard resulting from employment of personnel plus all overheads determined for a certain time period. This figure is reduced to an hourly rate and divided by the number of persons actually employed in boat construction to give the manhour cost rate. This is obviously somewhat higher than the take home hourly pay rate of the tradesman.

The steelwork which in weight and physical terms may constitute the greater part of the finished boat may in cost terms be as little as 25 percent of the total. This item is in effect the easiest to cost and the machinery and outfit which constitutes the bulk of the cost can prove more difficult and give rise to greater errors. A large proportion of this will need to be costed individually.

The major 'bought in' items should be costed by obtaining quotations from the manufacturers or his agent. The following would fall into this category:

(1) Main engine, shafting and propeller.
(2) Auxiliary machinery such as generators, pumps, power steering and refrigerating plant.
(3) Winches.
(4) Fishing gear such as power blocks, net drums, special davits, etc.
(5) Electronic devices such as radios, fishfinders, etc.

Installation of these items is costed on the basis of a knowledge of the material and labour costs involved in the installation of each item. Some of this installation work may be sub-contracted and a quotation from the sub-contractor is necessary.

The remaining general outfit of the boat may be divided into two groups: hull outfit and machinery outfit. For a first estimate of the hull and machinery outfit weights it is not uncommon for designers to employ the so-called Cubic Number (CUNO) principle whereby weights are compared on a volumetric basis for like hulls. The CUNO of a vessel is the product of multiplication of the vessels overall length, beam and moulded depth. For example the 15 m vessel has a CUNO of 15 x 5 x 2.4 = 180 and the 21 m vessel a CUNO of 21 x 6.5 x 3.6 = 491. To illustrate the method let us consider the hull outfit weight of the 15 m vessel which is 4.1 tons and use the CUNO ratio to estimate the hull outfit weight of the 21 m vessel. Thus,

$$4.1 \times \frac{491}{180} = 11.2 \text{ tons}$$

Actually the hull outfit weight of the 21 m vessel is 12.3 tons but there is a non-proportionately greater amount of fish hold insulation in the larger vessel which would have to be taken into consideration.

It is proposed in a number of publications that the subsequent costings of hull and machinery outfit be based on a cost per ton of outfit weight for labour and material. While this is sufficient for a preliminary estimate, in practice, where greater accuracy is required, this is difficult because many of the items normally associated with hull and machinery outfit have considerable variations in material costs and manhour rates.

For the hull outfit of a steel vessel, it is suggested that this be sub-divided into the four following categories for greater accuracy:

(1) Fish Hold Lining
(2) Accommodation outfit and miscellaneous woodwork
(3) Hardware - windows, vents, handrails, ladders, etc.
(4) Sandblasting, painting and anodes.

Given the costs for a similar vessel of like proportions the fish hold lining costs, both material and labour, may be compared in the ratio of the respective hold capacities, i.e. per cubic metre of fish hold. The accommodation outfit and hardware may be compared using the CUNO method and since paint costs are generally established on a cost per square metre basis, it is suggested that they be compared on a length squared (L^2) basis. In costing the labour any change in the hourly rate since building the previous vessel should be taken into account. Adjustment on a percentage basis should also be made for increases in material costs.

The machinery outfit can be sub-divided into the three following categories for greater accuracy.

(1) Engine related systems
(2) Pipe systems
(3) Electrical.

The first item which includes exhaust, controls, sea connections, lub oil tank and piping, is probably best calculated directly by individual component. It is difficult to relate it to engine size or cost and is independent of vessel size. As a preliminary estimate it may be 15 to 25 percent of the engine cost, the higher figure being for the cheaper model of engine. Pipe systems which include bilge, ballast and fire main, and general service systems, can be compared on a CUNO basis since they are a function of the size of the boat. The electrical system should be costed for the boat either by quotation if sub-contracted or by itemised material costs and manhour estimates if the boatyard is undertaking the work. Expensive items like switchboards should be costed by obtaining quotations.

To illustrate the method of cost estimating an example for the 15 m vessel follows. This example is for material and labour costs at the time of writing and for an experienced boatyard in an industrially developed country. Thus whilst the calculation represents a true situation it should only be treated as an example of the method and the figures given should not be used for an estimate at a later date or in another situation.

Cost Estimate for 15 m Steel Trawler

Item 1. Steelwork (Hull and Deckhouse)

Average steel costs

Mild steel plate	$	560/ton
Mild steel sections	$	645/ton
Tube and RHS material	$	850/ton

Weight of steel (tons) as per Tables 3 and 4:

	Net		Invoiced		
Plates	14.8	+	10%	=	16.28
Sections	5.2	+	5%	=	5.46

Of the sections it is estimated that 1.5 tons net is tube or RHS, i.e. 1.58 tons of invoiced steel, leaving 3.88 tons of other section steel.

Steel cost

Plate	16.28 tons x 560	=	$ 9,117
Sections	3.88 tons x 645	=	$ 2,503
Tube and RHS	1.58 tons x 850	=	$ 1,343
		Total	$ 12,963
Electrodes and consumables (+15%)		=	$ 1,943
		Total	$ 14,908

Labour (from Fig. 5) = 4200 manhours
at a charge rate of $11 per hour

	4200 x 11	=	$ 46,200
	Total for steelwork		$ 61,108

Item 2. Bought in items

(1) Main engine-quotation $ 24,500
    Installation materials and labour $ 7,000
    $ 31,500

(2) Shafting and propeller - quotation $ 10,000
    Installation materials and labour $ 1,600
    $ 11,600

(3) Motor Generator - quotation $ 3,000
    Installation materials and labour $ 800
    $ 3,800

(4) Steering Gear - quotation $ 2,500
    Installation material and labour $ 2,500
    $ 5,000

(5) Hydraulics for winch - quotation $ 1,800
    Installation material and labour $ 2,800
    $ 4,600

(6) Trawl winch - quotation $ 4,500
    Installation material and labour $ 850
                                                               $ 5,350

(7) Electronic instruments - quotation
                    Radio $ 2,000
                    Echo-sounder $ 1,500
                    Compass $ 400
    (includes installation cost)
                                                               $ 3,900

(8) Windlass - quotation $ 850
    Installation material and labour $ 450
                                                               $ 1,300

(9) Anchors and cables - quotation $ 1,200
    Installation material and labour $ 280
                                                               $ 1,480

              Total for bought in items $ 68,530

Item 3. Hull Outfit

Previous boat built of similar type, 22 m length
with CUNO of 396 and 80 m^3 fishhold capacity
had following costs in US dollars:

	Material	Labour
Fish Hold Lining	35,000	5,440
Accommodation outfit	12,700	25,080
Hardware	2,350	10,600
Painting and Anodes	5,800	5,950

If since building this boat material costs have risen
8% and the labour charge rate from $10 to $11 per hour
then for the 15 m boat, CUNO 180, fishhold 33 m^3.

Fishhold Lining

   Material = 35,000 × $\frac{33}{80}$ × 1.08   = $ 15,590

   Labour   = 5,440 × $\frac{33}{80}$ × $\frac{11}{10}$ = $ 2,470
                                                             $ 18,060

Accommodation Outfit

   Material = 12,700 × $\frac{180}{396}$ × 1.08   = $ 6,230

   Labour   = 25,080 × $\frac{180}{396}$ × $\frac{11}{10}$ = $ 12,540
                                                             $ 18,770

Hardware

   Material = 2,350 × $\frac{180}{396}$ × 1.08   = $ 1,150

   Labour   = 10,600 × $\frac{180}{396}$ × $\frac{11}{10}$ = $ 5,300
                                                             $ 6,450

Painting and anodes

    Material = 5,800 × $(\frac{15}{22})^2$ × 1.08 = $ 2,910

    Labour = 5,950 × $(\frac{15}{22})^2$ × $\frac{11}{10}$ = $ 3,040

                                                                $ 5,950

                Total for Hull Outfit $ 49,230

Item 4. Machinery Outfit

    Previous 22 m boat costs were in US$:

	Material	Labour
Pipe Systems	9,600	11,960

    Our pipe systems

        Material = 9,600 × $\frac{180}{396}$ × 1.08 = $ 4,710

        Labour = 11,960 × $\frac{180}{396}$ × $\frac{11}{10}$ = $ 5,980

                                                                     $ 10,690

    Engine related systems

        Say 20% of main engine
          quotation (24,500 × 0.2) = $ 4,900

    Electrical
        Quotation received = $ 6,000

                Total Machinery Outfit $ 21,590

Item 5. Miscellaneous Costs
(includes launching, insurance, trials,
surveys, etc.) $ 6,000

Item 6. Ballast $ 2,000

Item 7. Nozzle $ 5,000

                      GRAND TOTAL $213,458

                      Margin (5%) $ 10,672

                               TOTAL $224,130

ANNEX

OUTLINE SPECIFICATION FOR
15 m AND 21 m STEEL FISHING BOATS

The two designs illustrated in this handbook represent two quite different arrangements but having hulls which could be utilised with alternative layouts to suit a particular fishery. Basically both boats could be considered multi-purpose although the arrangements shown illustrate deck equipment for one particular type of fishing.

15 m Steel Fishing Boat

The design as presented in Drawing No. SB1-1 is purely for stern trawling and such a design may be utilised for one boat bottom trawling in coastal waters and both bottom and mid-water pair trawling. Where the fishing may be seasonal limiting the bottom trawling to a part of the year then it is useful if the vessel can be utilised without major modification to practice other forms of fishing. In this case using the basic hull design the deck equipment could be re-arranged to allow the vessel to be operated as a combination trawler/purse seiner or as a long liner/gill netter. Utilised solely as a trawler the stern gantry arrangement is preferred to carry the trawl blocks with a direct warp lead from winch and tackle for lifting the cod end over the transom. If the vessel were to be used as a combination trawler/purse seiner, the deck layout should be rearranged to provide a mast and boom for handling both the trawl cod end and the purse seine. Portable trawl davits should also be included so that the davit can be removed from the side on which the purse seine net is to be hauled. The offset wheelhouse is so placed to provide additional working space on the starboard side for either handling the purse seine or for siting a net/line hauler for line and gill net fishing. The hauler should be positioned forward where the helmsman can see the line or net coming aboard and there needs to be adequate room to work around the hauler. For purse seining the purse davit could be arranged on the starboard side with pursing wire leads from the athwartships winch drums to lead blocks at the rear of the house hence to the purse davit. (See FAO Fisheries Technical Paper No. 188, Fishing Boat Designs:3 - Small Trawlers). The house can be arranged on the centreline if the vessel is intended only for stern trawling or offset to starboard if the owner favours fishing from the port side.

The vessel has a large working deck area aft with raised forward deck for protection. An insulated fish hold for stowage of fish on ice is provided. With the insulated hold aft in this type of boat and to utilise fully the space, it will tend to have restricted depth aft and therefore two hatches are fitted to make loading and discharging easier and reduce wastage in warmer climates. Typical fish hatch construction is shown in Drawing No. SB1/2-4. The hold can usefully be subdivided for carrying ice to the grounds and smaller quantities of fish species if so desired. The hold in any case should be fitted with removable and semi-permanent fish hold divisions about 1.5 m apart. Access for inspection and re-packing of the stern gland is provided by a small section of portable insulation at the bottom of the fish hold in way of the shaft. It is preferred that lubrication of the bearing is done remotely from the engine room aft bulkhead area.

Accommodation consists of a wheelhouse with cooking and food storage/preparation facilities adequate for some three to four days at sea. A separate toilet/shower with entrance from the open deck is provided and there is below deck sleeping accommodation with four berths. A sliding door, shown on Drawing No. SB1-11, on the helmsman's side of the wheelhouse, does not interfere with the deck area and permits him to view any fishing operation on the deck in the vicinity of the wheelhouse. A good view of the winch and after working deck area which is a prerequisite of stern trawling operations is also afforded the helmsman by the wheelhouse location and windows at the back of the wheelhouse. The deckhouse construction is shown in Drawing No. SB1-7.

Deck gear consists of a two drum trawl winch with linked spooling equipment necessary for the direct warp lead to the stern gantry. The winch may be hydraulically driven with pump driven from the power take off on the main engine (or auxiliary engine) or can be mechanically driven with chain or belt drive from a pulley on the main engine power take off to a layshaft arrangement. A winch with a pull of 1 1/2 tons at mid drum and drum capacity of 1000 m of 12 mm diameter warp would be suitable for this particular boat. The anchor wire can be handled by the winch as shown on the general arrangement drawing, the wire being stowed on a separate reel. To aid the handling of the cod end and other fishing nets over the transom it is often good practice to fit a steel roller at the transom bulwark top. With a steel boat there is no reason why the upper rail of the transom should not consist of a heavy duty tube as presented in this design which serves the same purpose as a roller without its disadvantages. Maintenance of rollers and the snagging of nets at its support bearings are particularly troublesome. If purse seining operations are undertaken from the one side it is very important to ensure that the bulwarks on this side and transom are free of protuberances and sharp corners which can snag and tear nets. The stern gantry is arranged with the uppermost cross member supporting the cod end hoisting tackle inboard of the transom so that the cod end swings into the deck area for discharging when lifted above the transom. Retractable arms carrying the eyes for the trawl blocks are fitted to allow these to be stowed inboard of the belting line and the stern gantry carries a 1/2 ton SWL derrick for handling catches alongside or lifting gear. Provision can be made for adjusting the height of the trawl block for clearance of different trawl door sizes.

An economical heavy duty marine diesel engine has been shown in the design drawings but alternative main engines may be fitted in accordance with the owners and builders preference and availability. The Yanmar 6KD CGCE engine shown has a continuous rating of 165 hp at 1450 rpm and with a 3.55 to 1 reduction gearbox and 42" (1070 mm) diameter propeller in the nozzle gives an estimated thrust at 4 knots towing speed of almost 2.5 tons. With the nozzle installation in this boat a trawl of similar size to that for an installed power of 210-230 hp without nozzle can be handled with savings in investment and running costs.

Three tons of concrete ballast equally distributed fore and aft as shown in the General Arrangement plan is poured in the void spaces between floors after coating the steelwork with a bitumous paint. This seals these inaccessible areas and improves the vessels stability characteristics.

The rudder, stock and tube details are shown in Drawing SB1-10.

21 m Steel Fishing Boat

The arrangement for this vessel features accommodation and engine aft with a dry fish hold and three fish tanks forward as shown in Drawings Nos. SB2-1 and SB2-2. This type of arrangement is preferred in various areas for purse-seining and other types of fishing particularly in more exposed waters. The purse winch is located in the shelter of the forecastle and has adequate lead lengths to the purse davit located opposite on the starboard side well forward of amidships to aid manouvering when pursing and handling the net. Operations are easily observed from the bridge. A maximum clear working deck area is achieved in the forward part of the ship and the forecastle could be extended further aft on the port side to protect the winch and fish handling area from the sun or inclement weather if so desired. The open spaces at the sides of the deckhouse particularly on the starboard side and aft deck permit handling and stowage aft of the purse net. An after derrick boom stepped on the deckhouse top carries a purse block which is usually hydraulically driven to assist with net handling.

If so desired the boat may work as a stern trawler with suitable leads from the warp drums to blocks outboard then led aft to trawl blocks carried on arms built into the aft corners of the bridge deck. Positioning of the winch would have to be carefully considered to give the correct leads for trawling and purse seining if such a combination vessel is required. Also the vessel could be arranged for long lining with the line hauler mounted forward in the shelter of the forecastle.

The vessel as designed has three chilled sea water tanks and an ice hold. Alternative arrangements for the fish stowage space could be utilised to suit the owners preference. For the purse seining operation fish stowed in the chilled sea water tanks (50-60 percent of fish to 20-25 percent of sea water and 20-25 percent of ice) will maintain good quality. This is particularly important if food fish is to be the end product and it reduces the handling problem where large quantities of fish are required to be iced and stowed without crushing. Chilled sea water (CSW) as opposed to refrigerated sea water is proposed on the assumption that the time taken between taking on sea water after leaving port and beginning to stow the catch is insufficient to permit lowering the sea water temperature to the level required. This can be a problem particularly in areas where the initial sea water temperature is high and the fishing grounds not too distant from the point of loading. A refrigerated system could be installed for these tanks if the cost is justified and the system desired by the owner, the pump, valves and refrigeration plant being located in the engine room. A pump with sea suction and overboard discharge only is required for the chilled water system to fill and empty the tanks. Suctions from the tanks are fitted with strainers to prevent the size of fish caught being taken into the pump. Ice and bulk or boxed fish can be carried in the dry hold which is arranged with fish ponds. Usually compressed air is supplied to the CSW tanks to improve the cooling facilities.

Aft of the engine room there is a net and gear stowage space with hatch access from the main deck. This is to be lined out with timber sparring to suit the stowage requirements.

The forward mast carries a 1 ton SWL derrick which can be used for brailing and discharging iced fish. It could also be used to support a temporary canvas or similar awning over the foredeck if this was considered necessary. Forward of the dry fish hold bulkhead a void space aft of the fuel tank is provided for fitting an echosounder sonar transducer if required.

Accommodation consists of two deck cabins in the forecastle with up to four bunks in each. Deck cabins are preferred to below deck spaces for tropical conditions. The elevated wheelhouse has a clear view of the forward working deck and incorporates accommodation which could be occupied by the master if this is local practice. Windows in the after bulkheads and doors of this structure should be arranged to afford the helmsman a reasonable view aft particularly if trawling operations are planned. The main deckhouse contains a two berth cabin, mess and galley space for extended voyages and W/C and shower for the crew. The construction of the deckhouse is shown in Drawing No. SB2-7.

The engine shown is a Yanmar 6M-TE with continuous rating of 300 bhp at 750 rpm and a 2 to 1 reduction gearbox giving the vessel an estimated free running speed of 9.5 knots. If the vessel was to be used for trawling it is recommended that a nozzle and appropriate propeller is fitted.

The rudder, stock and tube details are shown in Drawing No. SB2-9.

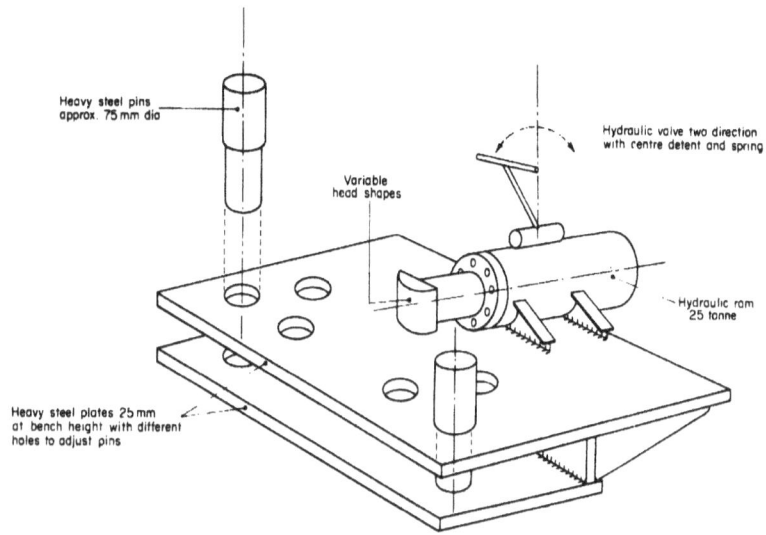

Fig. 1 BENDING PRESS FOR COLD BENDING FRAMES, STEM BARS ETC.

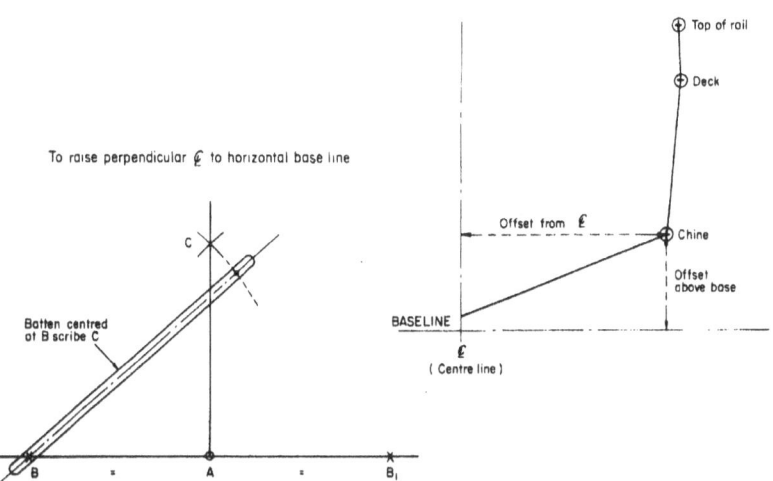

Fig. 2 LOFTING DETAILS

	Length of vessel	Weight of zinc anodes	Total No. of anodes
A	6 m - 12 m	8.6 kg	4
B	12 m - 15 m	8.6 kg	6
C	15 m - 18 m	13.1 kg	8
D	18 m - 21 m	15 kg	10
E	21 m - 24 m	15 kg	10
F	24 m - 27 m	15 kg	10

Blocks to be streamlined and have steel welding lugs cast into them

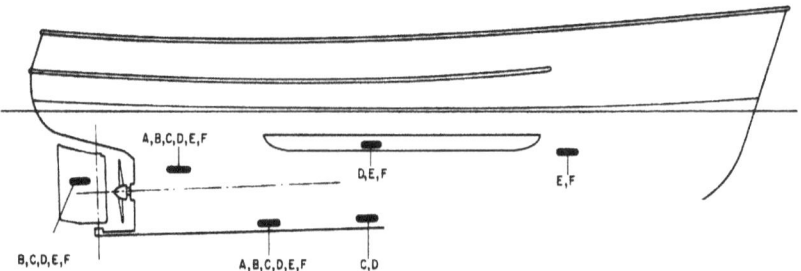

Note. Sacrificial anodes also fitted in way of sea water inlets if of non ferrous metal

Fig. 3 SACRIFICIAL ANODES GUIDE TO QUANTITY AND POSITION

Assumptions : 1) Similar proportions and scantlings to handbook designs
2) Vessels longitudinally framed.- If transversely framed reduce weight by 3 per cent.
3) Included are stern tube, rudder and tube, mast and derrick etc. Excluded are winch bases, gantries, nozzles, scrap ballast.

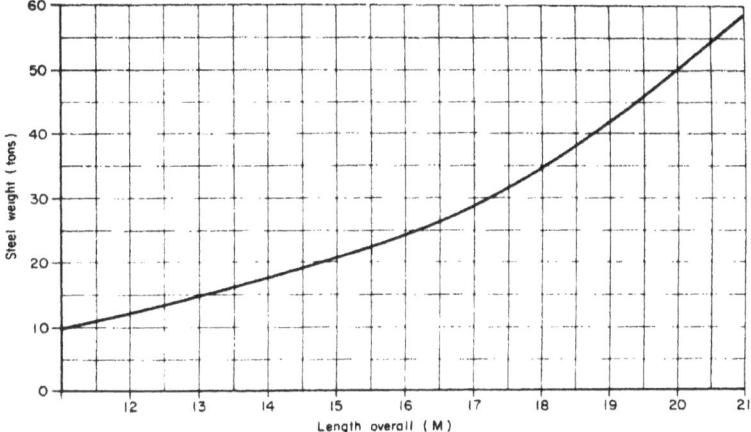

Fig. 4 NET STEEL WEIGHT

Notes. 1) Same assumptions are made as for net steel weight fig 4
2) Assumed longitudinal framing For transverse framing add 5 per cent total manhours

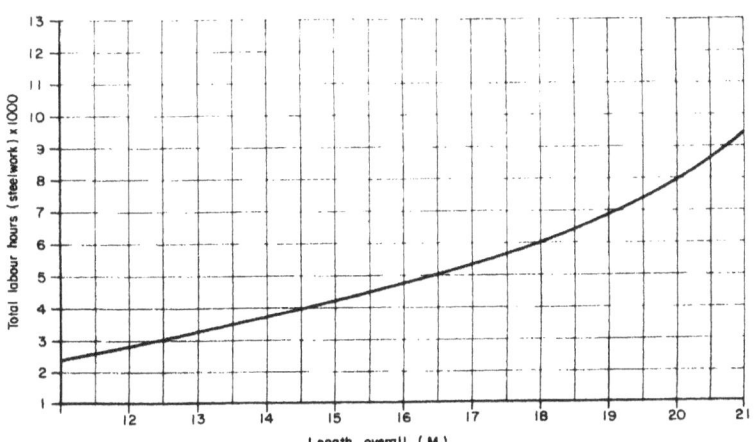

Fig. 5 MANHOURS (STEELWORK)

CONTINUOUS FILLET WELDS									
APPLICATION		PLATE THICKNESS (mm)	3	5	6	7	8	10	12
^^		TYPE OF RUN	LEG LENGTH (mm)						
Watertight bulkheads to hull tanksides and tankends to hull or bulkhead plating watertight or oiltight floor to hull plating.	Below top of floor line	Double heavy	5	5	5	5	5	6	7
Deckhouse front plating to deck. Mast and stays to doubling plates fittings to hull, deck or deckhouse, mast, boom, winch base subject to substancial stresses.		Single heavy	5	5	6	6	7	8	10
Watertight bulkhead to hull and deckplating tanktops and tankends or sides to hull or bulkhead plating.	Above top of floor line	Double medium	5	5	5	5	5	5	6
Deckhouse sides, aft ends to deck, hatch coamings and bulwarks to deck.		Single medium	5	5	5	5	5	6	7
Nonstructural connections requiring weather protection. Ventilators to deck, brackets for lights, buoys.		Light	5	5	5	5	5	5	5
INTERMITTENT FILLET WELDS									
Frames, floors, beams, stiffeners to hull, deck, deckhouse and bulkhead plating. (except as mentioned in notes below).		175mm distance staggered	5	5	6	6	7	8	10
Frames and floors to bottom plating in forward 1/3 part of vessel and inside tanks.		125mm distance staggered	5	5	6	6	7	8	10
Centre keelson to keel, keelsons to hull plating. Floors to keelsons. Floors in engine room to hull plating.		125mm distance chain	5	5	6	6	7	8	10

Notes:
if brackets, knees etc are fitted to the stiffeners, the weld is to be continued on both sides for the length of the bracket or knee.
If no knees or brackets are fitted, the weld is to be doubled for at least 150 mm length at the ends.

Staggered intermittent
| 150 mm | Distance C/C | 75mm |

Chain intermittent

Throat thickness not less than 0.70 x leg length unless otherwise mentioned

Plate thickness, for leg length calculation, to be assessed at not less than $\frac{A+B}{2}$

BUTT WELDS

APPLICATION	PLATE THICKNESS (mm)
Butts and seams of plating for hull, deck, deckhouse, bulkheads, tanks, bulwarks, platforms, insert plates. Use deep penetration electrodes only for the first run on these joints.	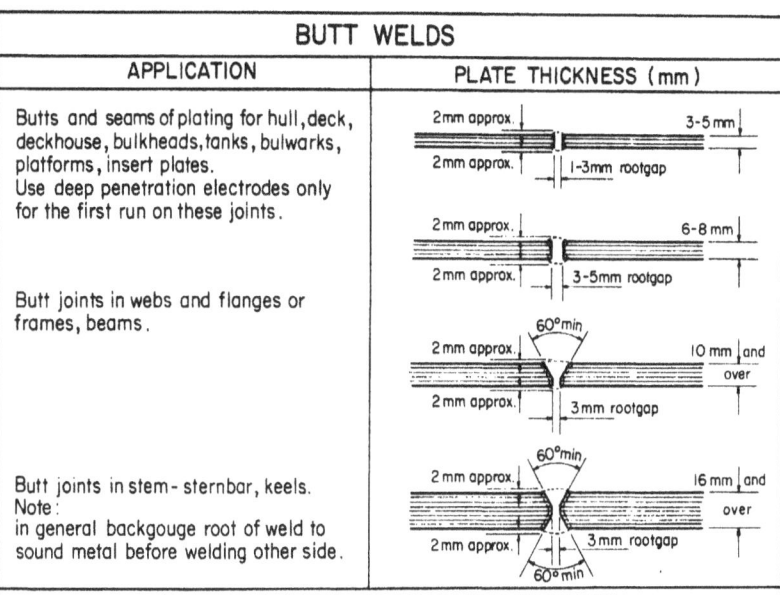
Butt joints in webs and flanges or frames, beams.	
Butt joints in stem - sternbar, keels. Note: in general backgouge root of weld to sound metal before welding other side.	

WELDING DETAILS OF MACHINE PARTS

Flanges on rudder stocks, tubes, compression flanges.	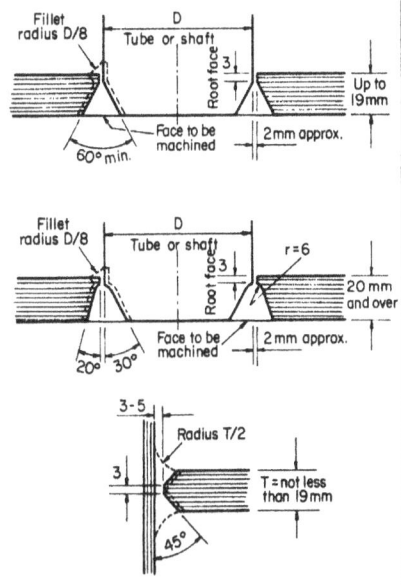
Annealing: welded machine components subject to heavy stresses are to be normalised (after welding) as follows :- A) components in the immediate vicinity of the welds are to be uniformly heated to a temperature of at least 1200°F (dull cherry red) B) components shall then be allowed to cool, covering them with fine dry sand or some other non-conductive material.	
Tiller arm connections to tiller head, stern post to stern boss.	

SPECIAL WELDS

APPLICATION	PLATE THICKNESS (mm)
BEVELLED FILLET WELD Connection of parts subject to heavy stresses or vibrations and where a double fillet weld is impossible or impracticable. Mast and stays to doubling plates.	45° min, 0-2 Leg length as per fillet tables 1-3
BUTT WELD DIFFERENT PLATE THICKNESSES Where plate thickness difference greater than 3mm chamfer thicker plate.	3mm 3mm root gap
PLUG WELD Inaccessible connections. Rudders, nozzles, tank tops, tank liners.	75 C/C distance as for intermittent fillet weld 2xt — Backing strip — Up to 12mm 2xt — Over 12mm
SLOTTED FILLET WELD Inaccessible connections.	75 2xt min — Backing strip Only for 12 mm and over

Other internal surfaces can be primed with traditional lead or aluminium based paints and finished with glass enamel paints.

TABLE 2
Typical Paint Systems

	Traditional System	One Component System	Two Component System
Hull	2-4 coats of underwater primer pigmented with aluminium and having lead content (1 coat being a shop primer) 2 coats of super antifouling	1 coat primer based on aluminium pigment and chlorinated rubber 2 coats chlorinated rubber paint 1 coat sealer 1 coat longlife antifouling	1 coat shop primer based on aluminium pigment 2 coats epoxy coal tar composition 1 coat sealer antifouling 1 coat longlife antifouling
Deck	2 coats primer with zinc chromate pigmentation 2 coats of deck paint	1 coat primer based on aluminium pigment and chlorinated rubber 1 coat chlorinated rubber paint 2 coats chlorinated rubber reinforced deck paint	1 coat shop primer based on aluminium pigment 2 coats epoxy coal tar composition 2 coats of deck paint
House	2 coats primer with zinc chromate pigmentation 1 coat Alkyd based undercoat 1 coat gloss synthetic enamel	1 coat primer as above 2 coats chlorinated rubber paint 1 coat chlorinated rubber finish paint	1 coat shop primer as above 2 coats epoxy coal tar composition 2 coats vinyl or chlorinated rubber finish paint

TABLE 3

STEEL QUANTITIES FOR 15m VESSEL

Mild Steel Plates (To B.S. 4360 ASTM-A131-74 or Equivalent)

No.	Item	Thickness mm	Area M²	Net Weight Kg.	Remarks
1	Rudder Top Flange	19	0.04	6.0	
2	Skeg Plate And Gussets	16	3.0	384.0	
3	Rudder Plate	13	0.93	93.0	
4	Rudder Stiffeners	13	0.25	25.0	
5	Gussets	12	1.0	100.0	
6	Anchor Roller Cheek Plates	12	0.6	60.0	
7	Engine Bed Vertical Plate	10	3.3	264.0	
8	Hatch Fitting	10	0.5	40.0	
9	Floors, Lower Bulkhead Strakes, Knees	8	26.5	1696.0	Deckbeam and Chine Knees
10	Engine Bed End Gussets	8	0.75	48.0	
11	Tanksides	8	4.5	288.0	200 mm Outboard of centre
12	" (D/E-I)	8	4.0	256.0	1350 mm Outboard of centre
13	Deck Inserts	8	1.0	64.0	
14	Nozzle Cylindrical Plate	8	1.2	76.8	
15	Tanksides (G-I)	6	3.5	168.0	) 1350 m Outboard of
16	" (T-B)	6	3.1	148.8	) Centre. Net Figure) includes Cutouts for) Manholes
17	Tanktop (D/E-G)	6	8.0	384.0	
18	Tanktop (G-I)	6	4.7	225.6	
19	Manhole Plates	6	2.0	96.0	
20	Tank Ends (I) and Baffles (G-H) and (H-I)	6	3.9	187.2	
21	Bulkhead Plating above Floors	6	25.7	1233.6	
22	Inspection Plate over Tunnel	6	0.5	24.0	
23	Bottom Plating	6	63.0	3024.0	
24	Nozzle Conical Plate + Stiffs.	6	3.0	144.0	
25	Deck Plating	5	62.0	2480.0	
26	Side Plating and Fwd Bulkheads	5	52.0	2080.0	
27	Transom Plating	5	3.9	156.0	
28	Hatch Coamings	5	4.7	188.0	
29	Hatch Covers	5	3.2	128.0	
30	Nozzle Throat Plate	5	0.9	36.0	
31	Bulwark Plating	4	19.0	608.0	Freeing Ports Included
	Total			14712.0	

Note   Plate Material Listed is Nett Quantity
        Add 10% Waste Allowance

TABLE 4

STEEL QUANTITIES FOR 15m Vessel

Mild Steel Sections

No.	Item	Section Size (mm)	Length (m)	Weight (kg)	Remarks
1	Keel and Lower Stem Bar	150 x 16	7.7	145.0	Flat Bar
2	Upper Stem Bar	120 x 16	2.7	40.7	" "
3	Skeg Bottom Reinforcing	40 x 25	4.5	35.3	" "
4	Sole Piece	120 x 50	2.2	103.6	" "
5	Stern Post	120 x 60	1.4	79.2	" "
6	Stern Bar	120 x 16	1.8	27.2	" "
7	Transom C.L. Stiffener	100 x 16	1.0	12.6	" "
8	Flanges on Floors	75 x 8	36.0	169.6	" "
9	Flanges on Eng. Bed Gussets	80 x 8	2.9	14.6	" "
10	Engine Bed Top Plates	150 x 19	5.8	136.6	" "
11	Girder (G-J) Web	140 x 10	7.1	78.0	" "
12	" " Face Bar	100 x 12	7.1	66.9	" "
13	" (J-L) Web	100 x 6	2.2	10.4	" "
14	" " Face Bar	100 x 10	2.2	17.3	" "
15	Bulwark Stiffeners	100 x 8	18.6	116.8	" "
16	Hatch Edge Stiffening	60 x 6	14.0	39.6	" "
17	" " "	20 x 6	14.0	13.3	" "
18	Rail Stanchions	50 x 8	3.5	11.0	" "
19	Chine Bars	50 x 12	30.0	141.3	" "
20	Transverse Frames	100 x 75 x 8	28.3	298.3	Angle Bar
21	Transverse Beams	100 x 75 x 8	34.7	365.7	" "
22	Engine Bed Vertical Stiffs.	100 x 75 x 8	2.9	30.6	" "
23	Manhole Frames	50 x 50 x 6	11.0	49.2	" "
24	Bulkhead Stiffeners	40 x 40 x 6	64.0	225.3	" "
25	Girder (E-G)	100 x 75 x 8	4.8	50.6	" "
26	Deck Reinforcing	100 x 75 x 8	3.3	34.8	" "
27	Tank Top Stiffeners	40 x 40 x 6	14.0	49.3	" "
28	Deck Longitudinals	40 x 40 x 6	132.0	464.6	" "
29	Hull Longitudinals	40 x 40 x 6	184.0	647.7	" "
30	Transom Stiffeners	40 x 40 x 6	6.0	21.1	" "
31	Hatch Stiffeners	75 x 50 x 6	3.0	16.9	" "
32	Stern and Gland Boss	236 O/D x 140 I/D	0.6	106.8	Mild Steel Tube
33	Stern Tube	150 O/D x 118 I/D	2.5	143.5	" " "

Mild Steel Sections (Continued)

No.	Item	Section Size (mm)	Length (m)	Weight (kg)	Remarks
34	Rudder Tube	100 O/D x 74 I/D	0.4	11.6	Mild Steel Tube
35	Bulwark Rail	60 O/D x 40 I/D	31.4	486.7	" " "
36	Transom Bulwark Rail	150 O/D x 125 I/D	4.3	202.5	" " "
37	Bollards	125 O/D x 105 I/D	1.6	60.5	" " "
38	Rudder Tube Ends	100 Dia	0.25	18.7	Solid Round Bar
39	Fairleads	25 Dia	1.3	5.0	" " "
40	Lower Rails fo'c's'le Deck	19 Dia	9.5	21.2	" " "
41	Rudder Main Piece	50 Dia	1.3	20.0	" " "
42	Nozzle Trailing Round	20 Dia	3.5	7.8	" " "
43	Nozzle Leading Round	35 Dia	4.2	31.7	" " "
44	Pillars	50 x 50 x 6	11.8	57.5	Rectangular Hollow Section
45	Belting (to be split in halves)	150 x 75 x 10	14.0	429.8	" " "
46	Top Rail fo'c's'le Deck	40 N	9.5	47.5	Steel Pipe
			Total Weight	5163.9	

Note Section Material Listed is Nett Quantity Add 5% Waste Allowance

TABLE 5

STEEL QUANTITIES FOR 21m VESSEL

Mild Steel Plates (To BS4360, ASTM-A131-74 or Equivalent)

No.	Item	Thickness mm	Area M²	Net Weight Kg.	Remarks
1	Sternframe Gussets	25	0.5	100.0	
2	Rudder Top and Bottom Flange	19	0.4	60.8	
3	Anchor Roller Cheek Plates	16	0.5	64.0	
4	Rudder Plate	16	2.0	256.0	
5	Rudder Stiffeners	16	0.5	64.0	
6	Gussets	16	0.2	25.6	
7	Engine Bed Vertical Plates	10	6.5	520.0	
8	Deck Insert Plates	10	2.0	160.0	
9	Hatch Fittings	10	0.7	56.0	
10	Centre Keelson and Gusset	8	9.1	582.4	
11	Floors, Lower Bhd Strakes and Knees	8	49.0	3136.0	Including Deckbeam and Chine Knees
12	Brackets on Eng. Bed Plates	8	1.0	64.0	
13	Tank Sides (T-B)	8	5.8	371.2	) Includes Cut Outs
14	" " (E-H)	8	9.8	627.2	) for Manholes
15	Tank Margin Plates (J/K-M)	8	2.0	128.0	
16	Tank Tops (E-H)	8	10.0	640.0	
17	Manhole Cover Plates	8	3.6	230.4	
18	Skeg Closing Plate	8	0.8	51.2	
19	Tank Ends (E) and Baffles (A) and (F) (G)	8	10.0	640.0	
20	Main Deck Stringer Plate	8	11.6	742.4	
21	Bottom Plating	8	125.9	7050.9	
22	Tank Top (J/K-M)	7	9.0	504.0	
23	Tank Top (M/N-O)	7	6.7	375.2	
24	Tank Ends (J/K) and (M/N) and Baffle (N)	7	28.3	1584.8	Above Floor Plates
25	Bulkhead Plating	7	67.0	3752.0	" " "
26	Main Deck Plating	7	101.1	5661.6	Hatch Openings Excluded
27	Fo'c's'le Deck Plating	7	18.7	1047.2	
28	Hull Side Plating	7	106.3	5952.8	

Mild Steel Plates (Continued)

No.	Item	Thickness mm	Area M²	Net Weight Kg.	Remarks
29	Transom Plating	7	6.8	380.8	
30	Hatch Coamings	7	9.6	537.6	
31	Hatch Covers	7	7.5	420.0	
32	Centre Tank Division (M/N-O)	6	3.5	168.0	
33	Tank Liners	6	93.5	4612.0	Including Inside Hatch Coamings
34	Bulwark and Whaleback Plating	6	44.7	2145.6	Freeing Ports Included
35	House Front Plating Lower	6	7.5	360.0	*
36	Longitudinal Bulkhead (M-O)	5	5.5	220.0	
37	House Front Plating - Wheelhouse	5	6.0	240.0)	* Includes Cut-Outs for Windows
38	House Side Plating Lower	5	17.5	700.0)	but Excludes
39	House Rear Plating Lower	5	6.0	240.0)	Door Openings
40	Plating D-D	5	2.2	88.0)	
41	Top of Lower House	5	21.3	852.0	
42	Wheelhouse Side Plating	4	4.2	134.4)*	
43	Wheelhouse Rear Plating	4	5.0	160.0)	
44	Deck Locker	4	6.7	214.4	
45	Top of Wheelhouse	4	9.3	297.6	
46	Steel Doors	4	1.8	57.6	
47	Internal Partitions	3	13.8	331.2 *	
48	Vent Trunks and Cowls	3	3.4	81.6	
	Total			46688.5	

Note   Plate Material Listed is Nett Quantity Add 10% Waste Allowance

TABLE 6

STEEL QUANTITIES FOR 21m VESSEL

Mild Steel Sections

No.	Item	Section Size (mm)	Length (M)	Weight (Kg)	Remarks	
1	Keel and Lower Stem Bar	220 x 25	18.8	846.0	Flat	Bar
2	Stem Bar	150 x 25	3.9	115.0	"	"
3	Stern Bar	150 x 25	2.6	76.5	"	"
4	Engine Bed Horizontal Plates	150 x 25	10.3	303.2	"	"
5	Sole Piece	150 x 60	3.0	222.0	"	"
6	Stern Post	150 x 60	2.2	155.0	"	"
7	Transom and Reinforcing	100 x 16	1.3	16.3	"	"
8	Stiffeners on Keelson	80 x 8	7.7	38.7	"	"
9	Flanges on Floors and Knees	80 x 8	66.8	335.0	"	"
10	Flanges on Engine Bed Brackets	80 x 8	3.2	16.06	"-	"
11	Flush Manhole Frames	65 x 16	2.0	16.3	"	"
12	Flat Bars on Fishhold Pillars	38 x 5	80.0	119.2	"	"
13	Bulwark Stanchions	120 x 10	27.5	259.0	"	"
14	Hatch Edge Stiffening	60 x 6	21.0	59.4	"	"
15	Margin Bar for Deckhouse Lining	60 x 6	50.0	141.5	"	"
16	Hatch Edge Stiffening	20 x 6	21.0	20.0	"	"
17	Rail Stanchions	50 x 8	43.8	137.5	"	"
18	Web Girder Deckhouse Top	50 x 8	1.6	5.0	"	"
19	Stiles Between Windows	50 x 8	6.5	20.4	"	"
20	Chine Bars	65 x 12	42.0	257.0	"	"
21	Edge Bar of Wheelhouse Top	80 x 6	14.3	53.9	"	"
22	Edge Bar of Deckhouse Top	80 x 6	15.2	57.3	"	"
23	Face Bar of Wheelhouse Girder	80 x 10	2.2	13.8	"	"
24	Face Bar of Deckhouse Girder	20 x 10	1.6	10.0	"	"
25	Girder in Deckhouse	60 x 10	0.75	3.5	"	"
26	Door Frame	50 x 6	8.5	20.0	"	"
27	Door Edge Bars	50 x 6	8.5	20.0	"	"
28	" " "	20 x 5	8.2	6.5	"	"
29	Transverse Frames	120 x 80 x 8	54.4	663.7	Angle Bar	

Mild Steel Sections (Continued)

No.	Item	Section Size (mm)	Length (M)	Weight (Kg)	Remarks
30	Transverse Beams	120 x 80 x 8	60.9	743.0	Angle Bar
31	Engine Bed Vertical Stiffeners	120 x 80 x 8	4.7	57.3	" "
32	Stiffeners for Tanks	120 x 80 x 8	23.2	283.0	" "
33	Girders (K-M)	120 x 80 x 8	5.1	62.2	" "
34	Tank Side and Top Stiffeners	50 x 50 x 8	83.6	486.6	" "
35	Stiffeners Tank Ends and Baffles	50 x 50 x 8	65.0	378.3	" "
36	Bulkhead Stiffeners	50 x 50 x 8	169.5	986.5	" "
37	Main Deck Longitudinals	50 x 50 x 8	200.0	1164.0	" "
38	Fo'c's'le Deck Longitudinals	50 x 50 x 8	43.4	252.6	" "
39	Hull Longitudinals	50 x 50 x 8	425.0	2473.5	" "
40	Transom Stiffeners	50 x 50 x 8	12.5	72.8	" "
41	Manhole Frames	50 x 50 x 6	21.6	96.6	" "
42	Lugs for Tank Liners	100 x 75 x 7	27.5	256.3	" "
43	" " " "	150 x 80 x 8	3.0	45.9	" "
44	Girders (D-H)	150 x 75 x 11	10.2	189.7	" "
45	Hatch Stiffeners	75 x 50 x 6	3.5	19.7	" "
46	Deckbeams - Deckhouse and Wheelhouse	40 x 40 x 6	54.0	190.0	" "
47	Deckhouse Stiffeners	40 x 40 x 6	34.0	119.7	" "
48	Wheelhouse Stiffeners	40 x 40 x 6	64.0	225.3	" "
49	Stern and Gland Boss	273 OD x 173 ID	0.7	191.0	Steel Tube
50	Stern Tube	180 OD x 140 ID	1.3	110.9	" "
51	Rudder Tube	140 OD x 100 ID	0.7	44.0	" "
52	Rudder Tube Ends	140 OD x 80 ID	0.4	33.8	" "
53	Pillars	65 OD x 53 ID	6.9	69.0	" "
54	Rail Bulwark Cap/Fender	65 OD x 45 ID	54.0	696.0	" "
55	Bollards	125 OD x 105 ID	1.6	59.2	" "

Mild Steel Sections (Continued)

No.	Item	Section Size (mm)	Length (M)	Weight (Kg)	Remarks
56	Towing Post	125 OD x 105 ID	2.6	96.2	Steel Tube
57	Pillars in Fishhold	50 x 50 x 6	10.4	69.1	Rectangular Hollow Section
58	Belting (to be split in halves)	200 x 100 x 10	40.0	1880.0	" " "
59	Girder in Wheelhouse	100 x 50 x 4	2.2	20.8	" " "
60	Fairleads	25 ø	1.3	5.0	Solid Round Bar
61	Guard Rails	19 ø	63.4	141.4	" " "
62	Rudder Main Piece	70 ø	2.0	60.4	" " "
63	Top of Fo'c's'le Rail	BSP 40 NOM	16.5	165.0	Steel Pipe
64	Top of House Rail	BSP 40 NOM	15.2	76.0	" "
65	Lugs for Tank Liners	30 x 30 x 5	13.0	26.0	Tee Bar
		Total		15854.56	

Note   Section Material Listed is Nett Quantity Add 5% Waste Allowance

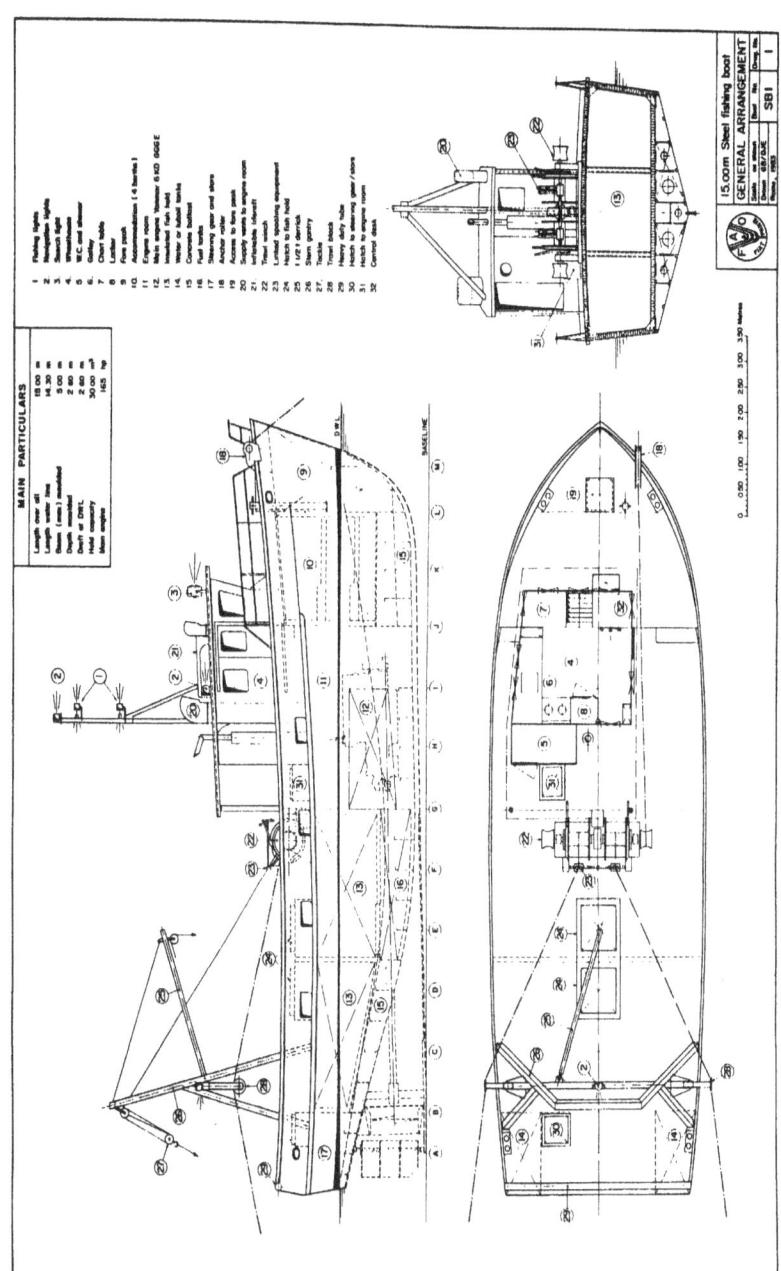

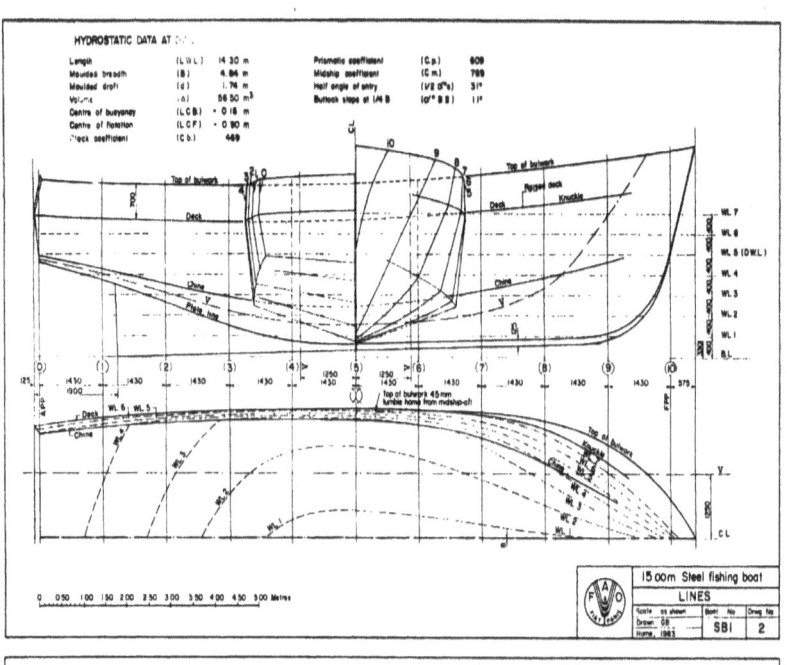

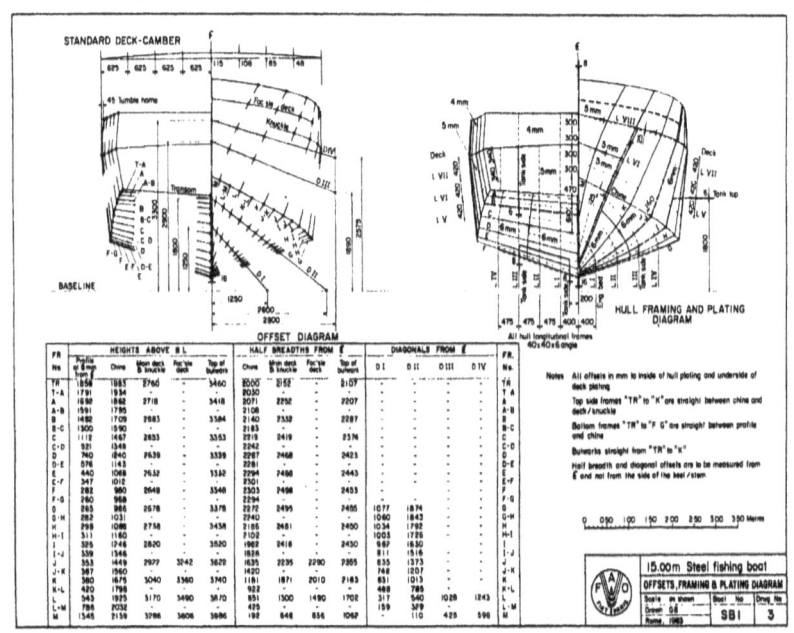

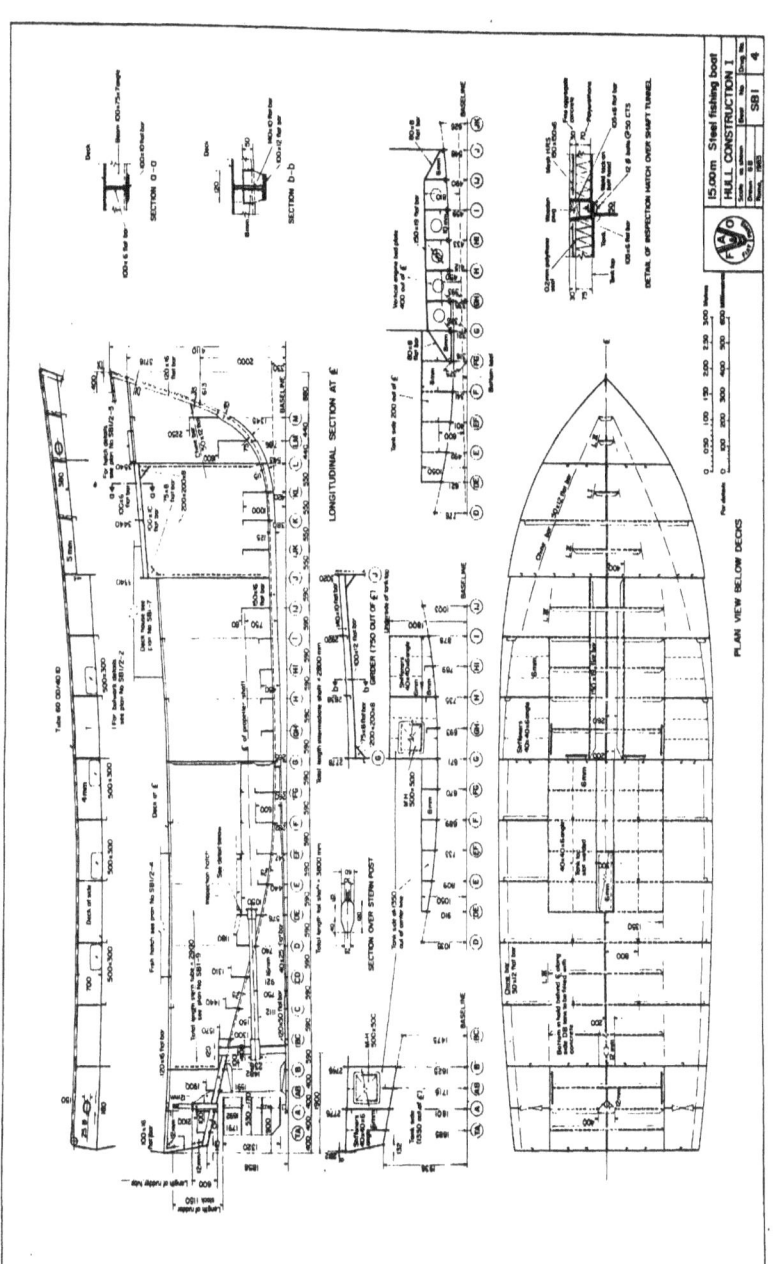

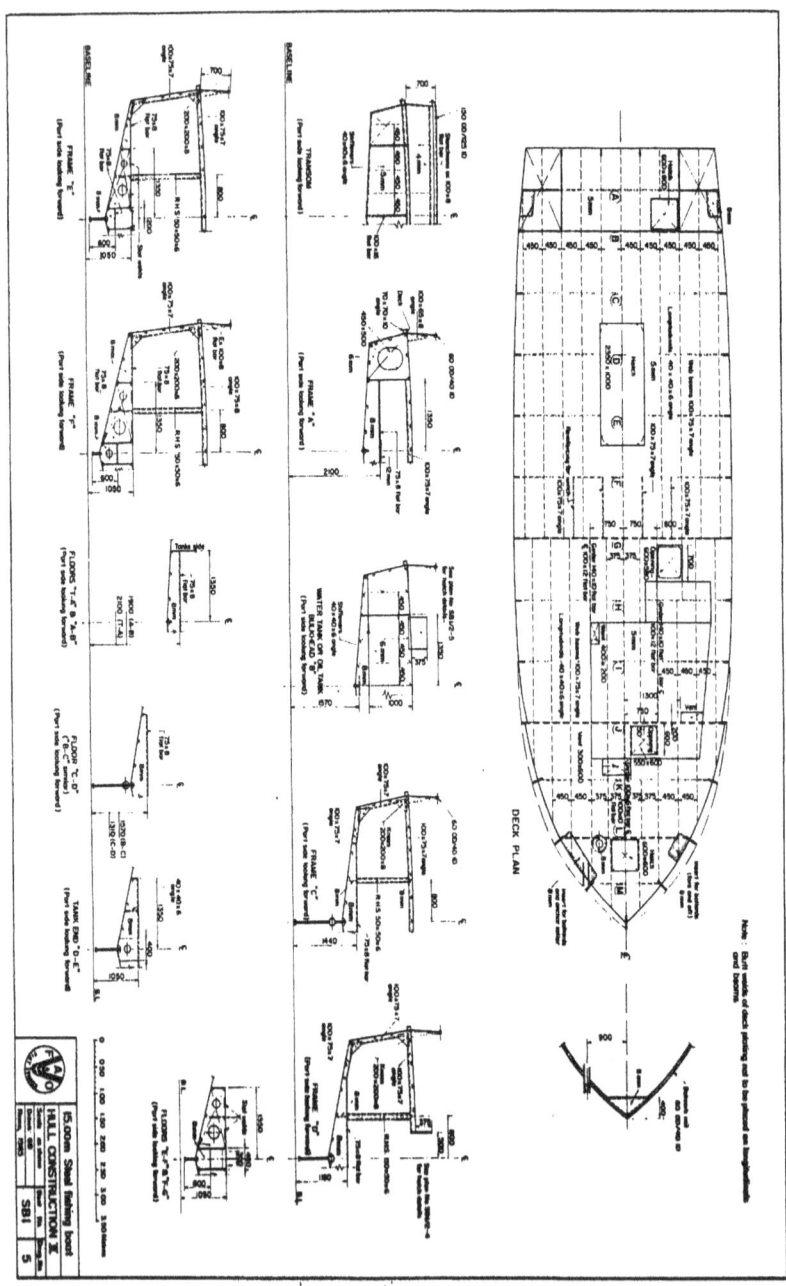

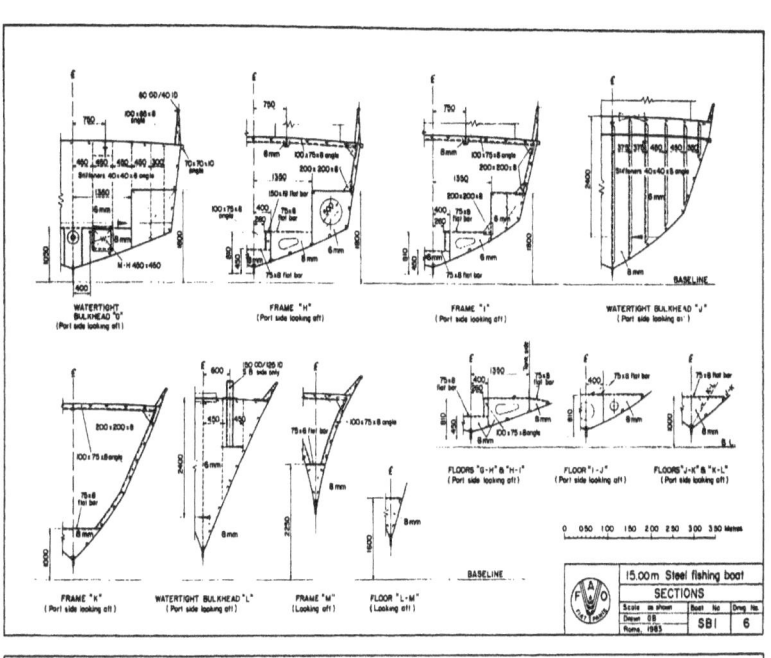
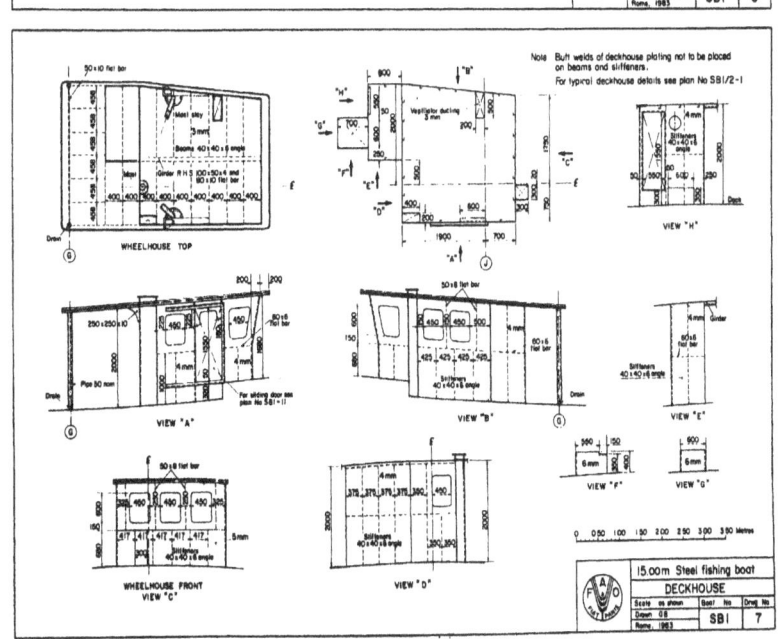

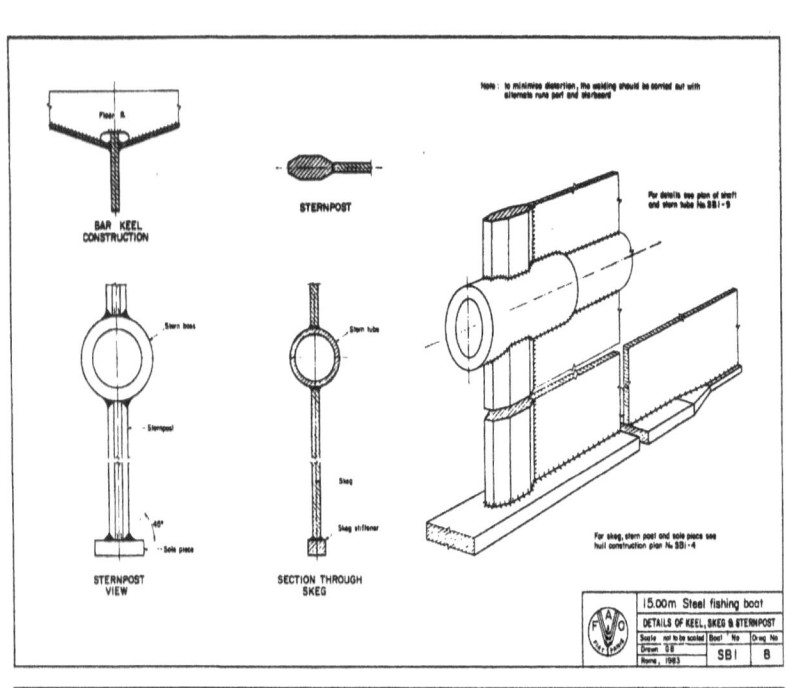

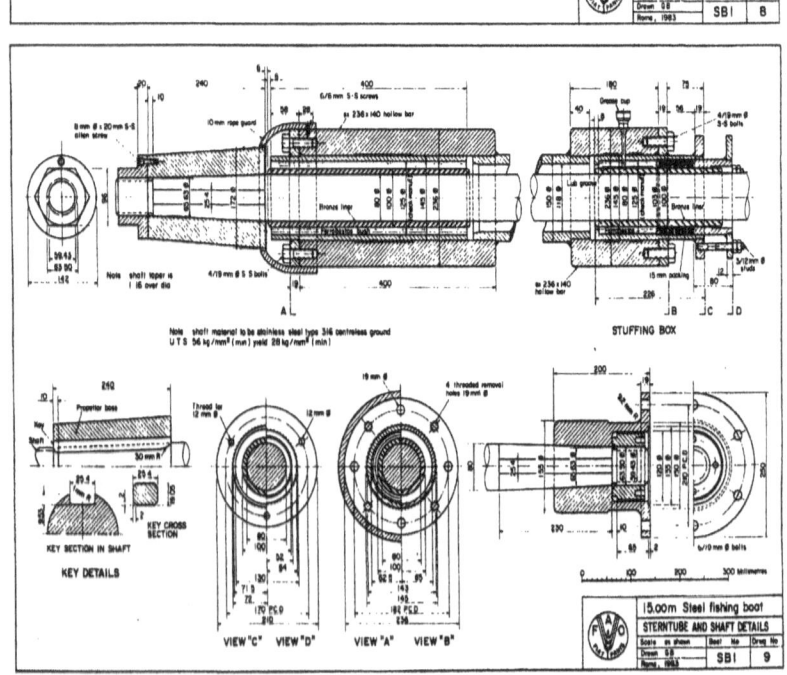

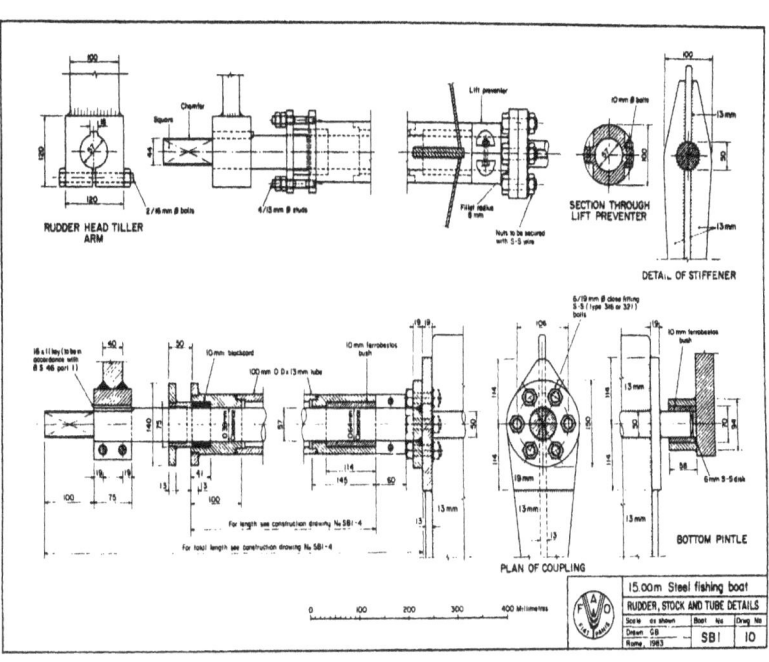

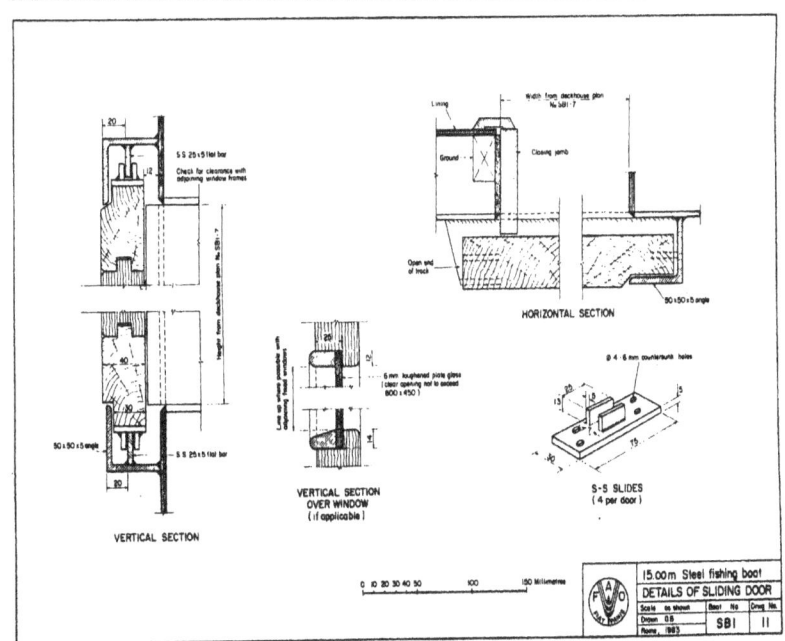

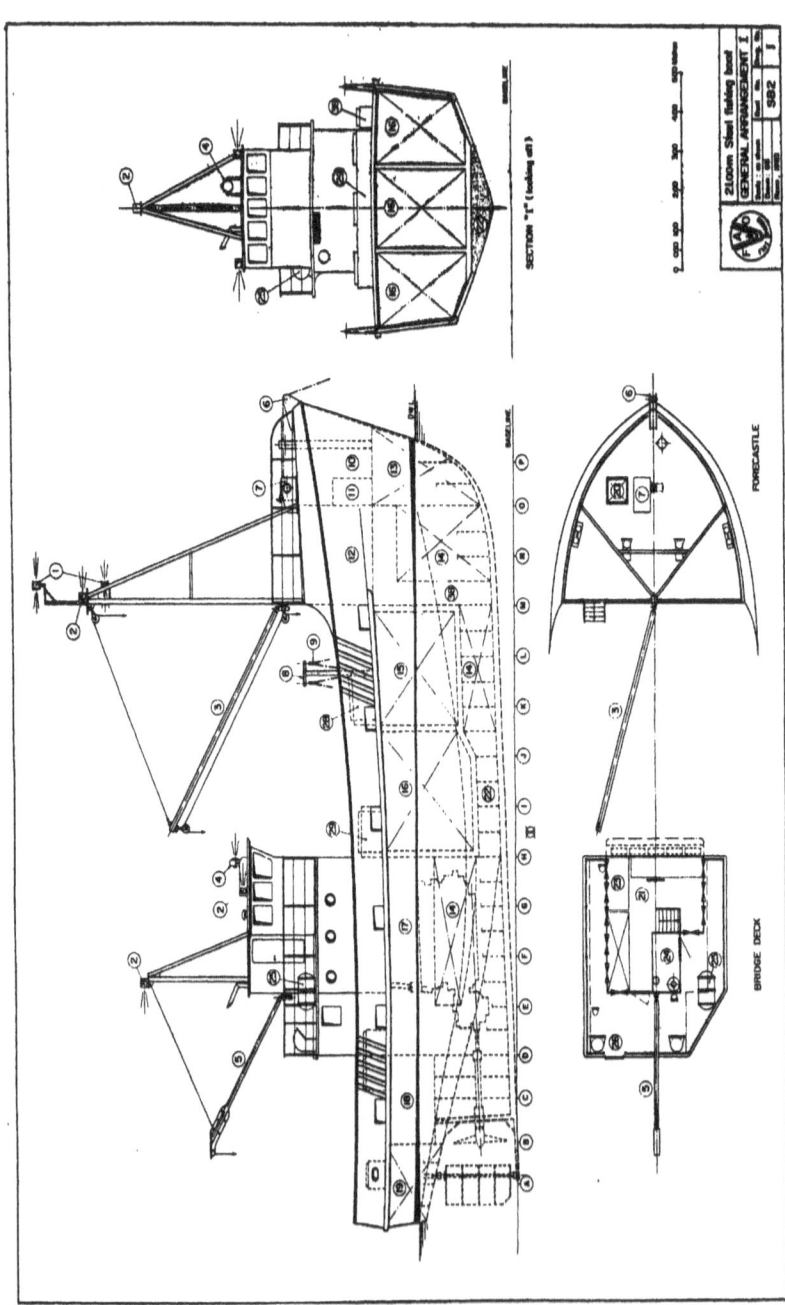

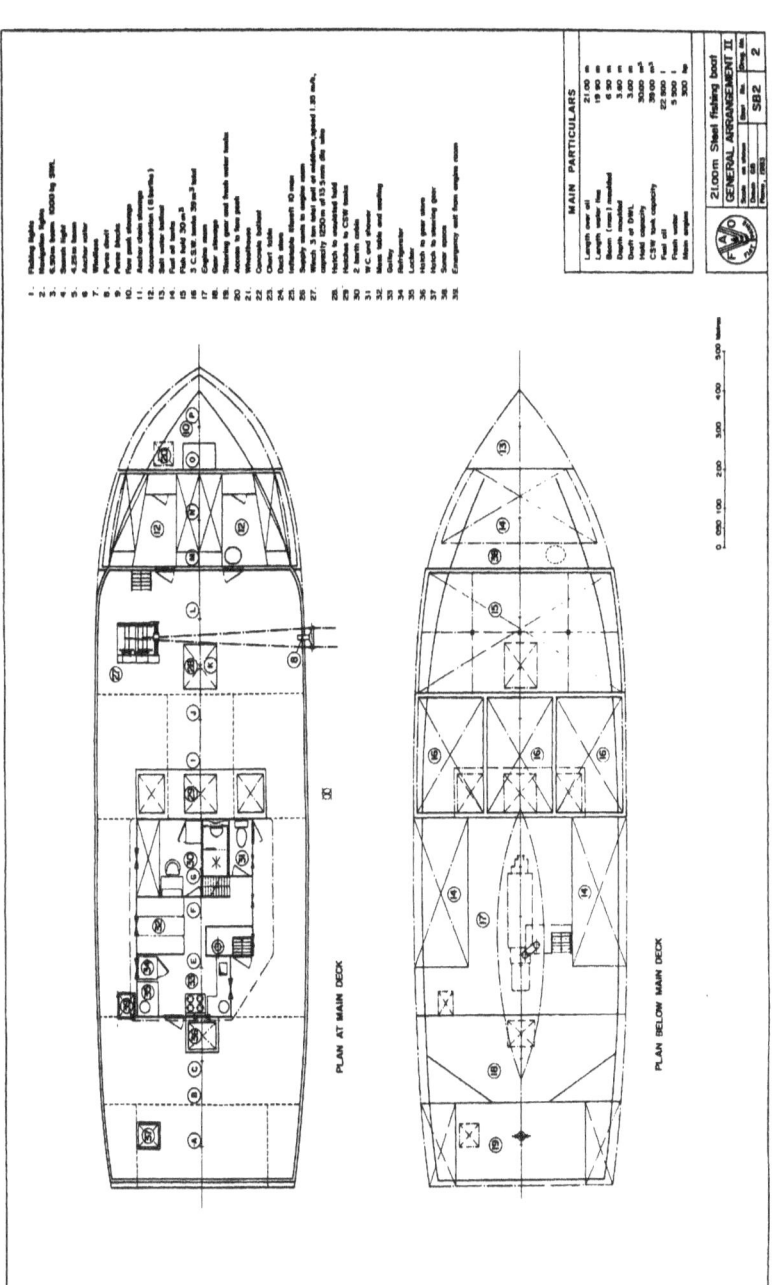

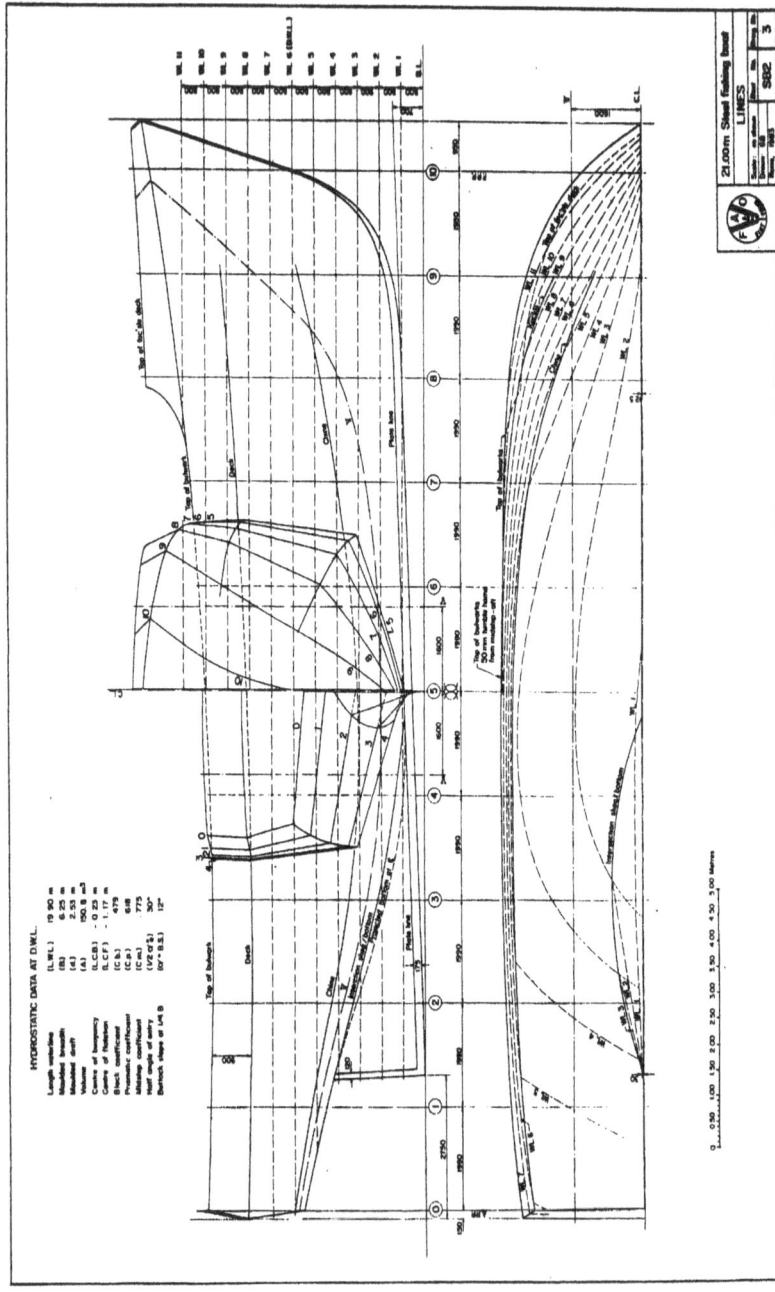

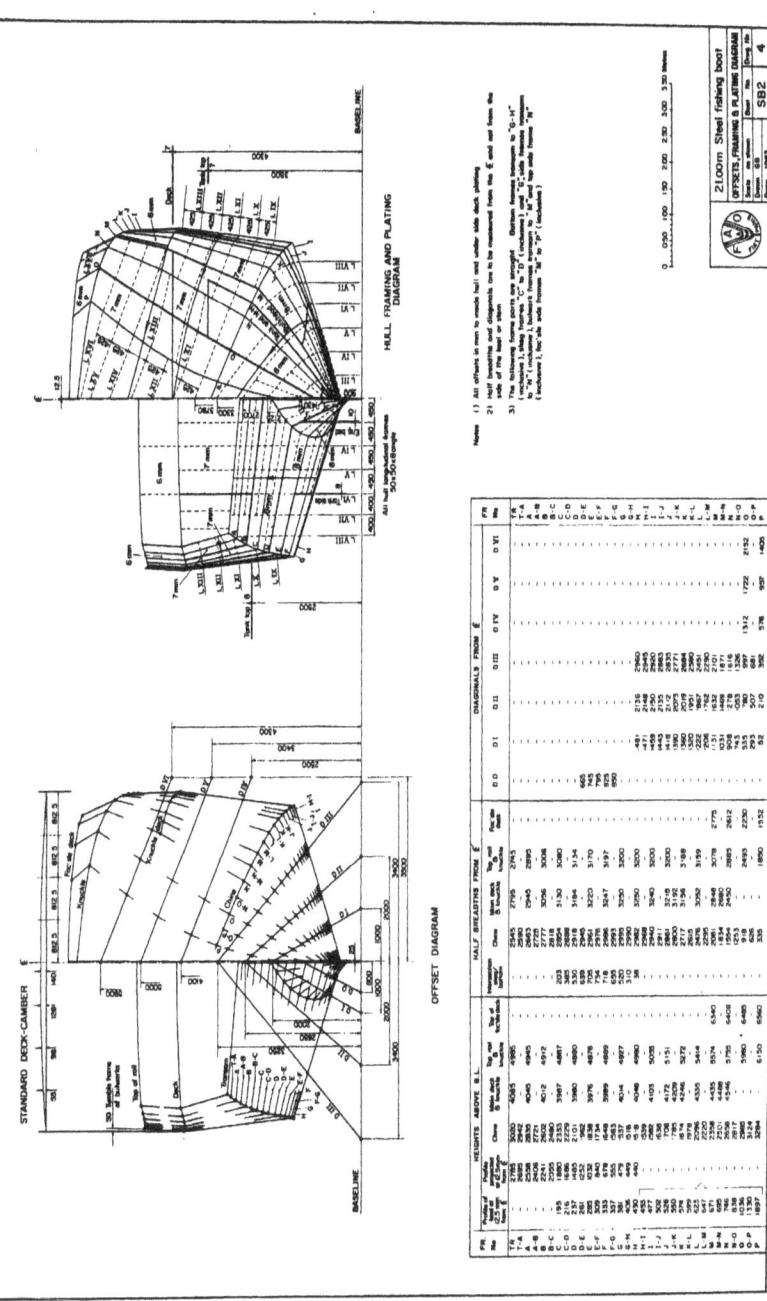

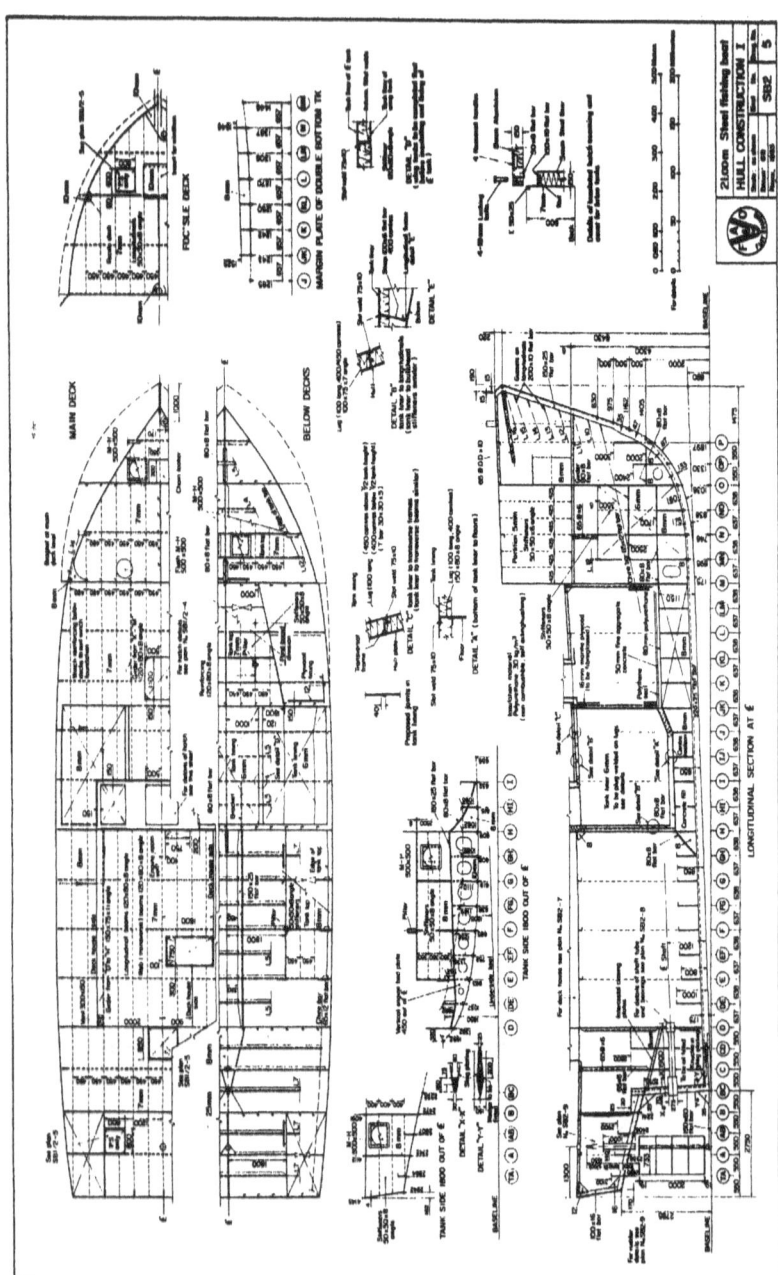

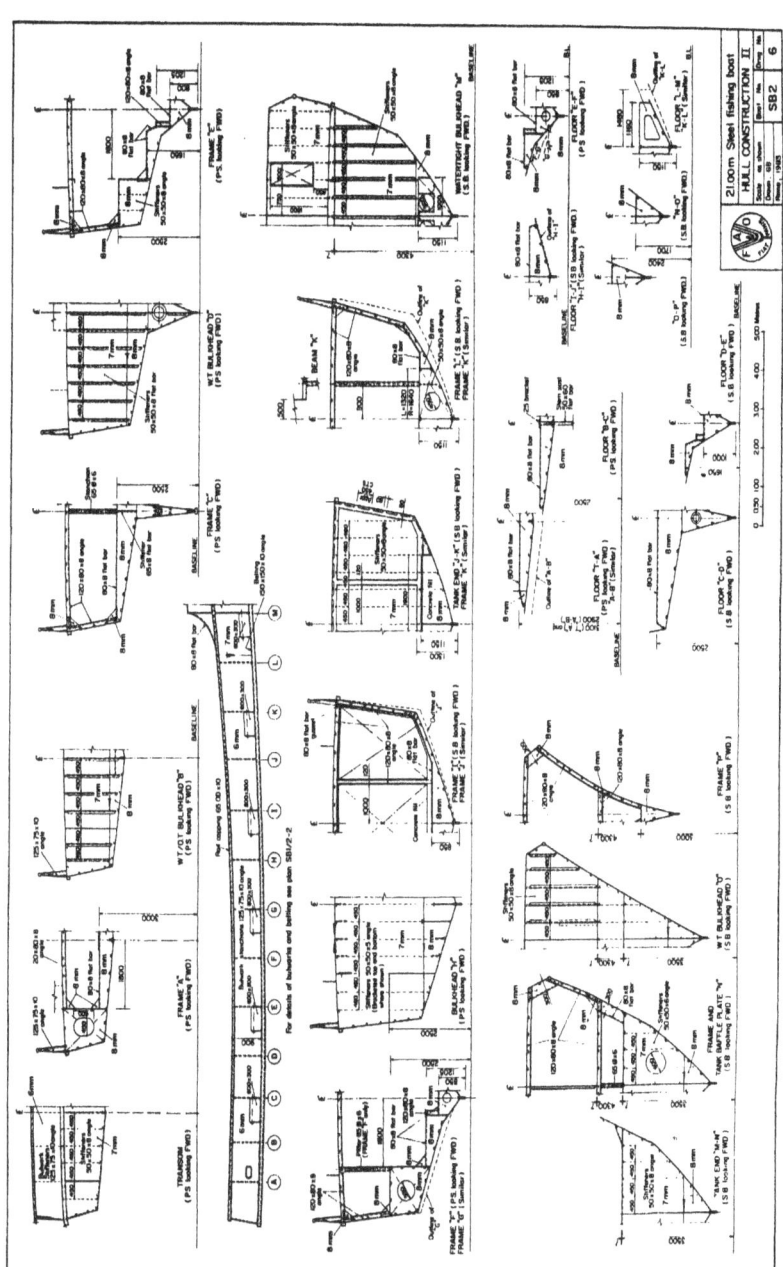

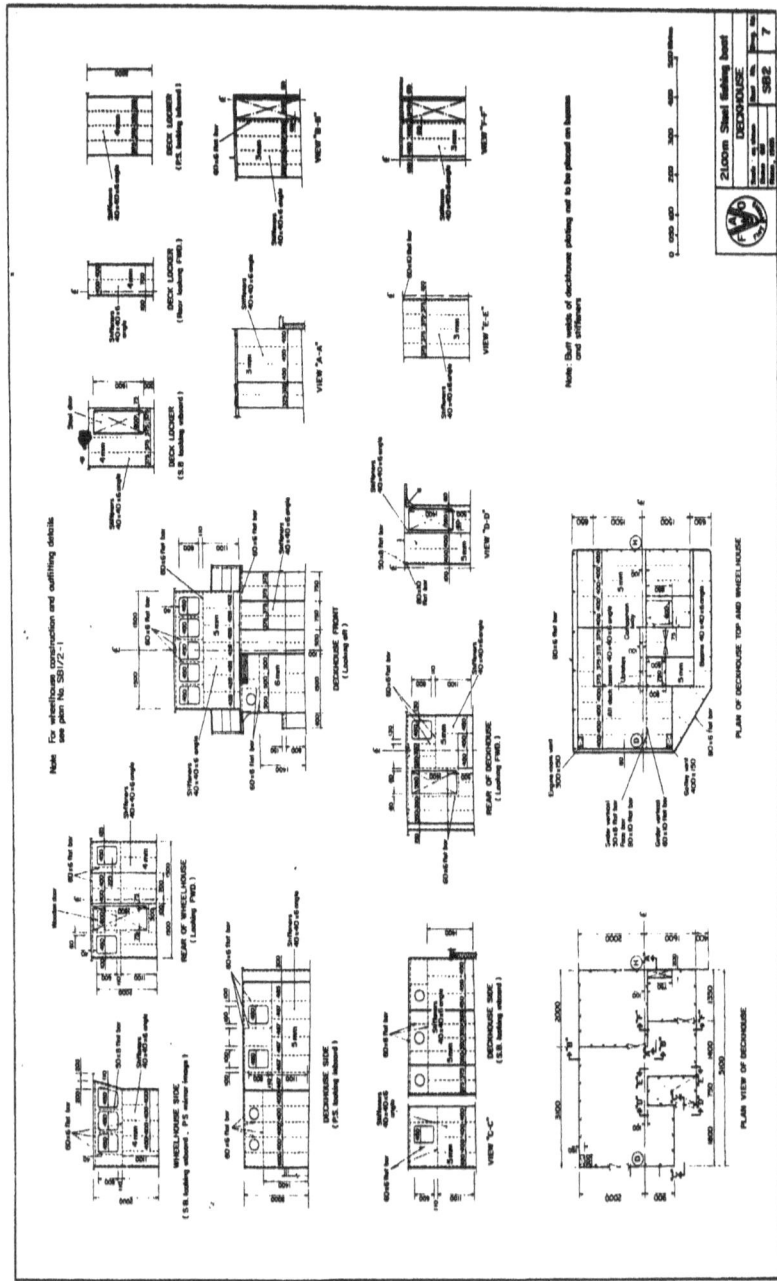

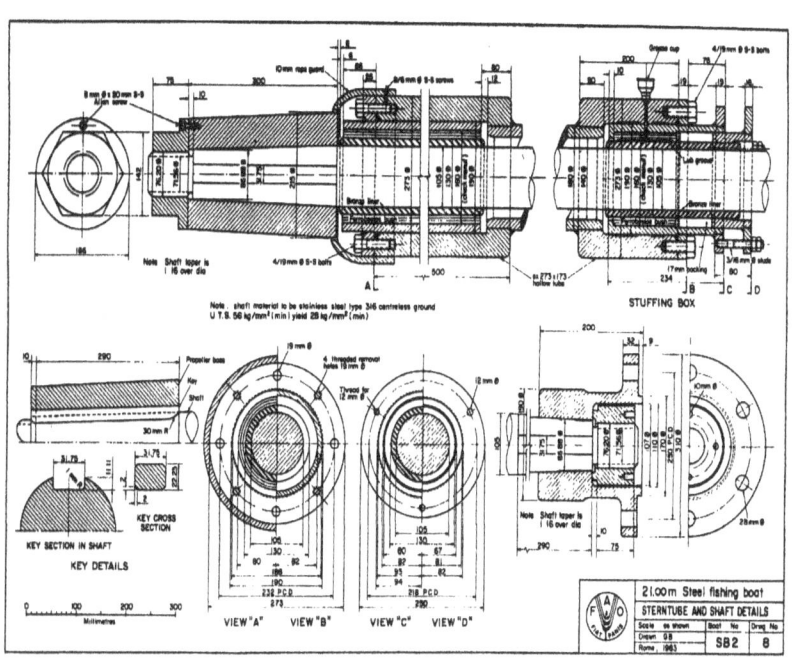
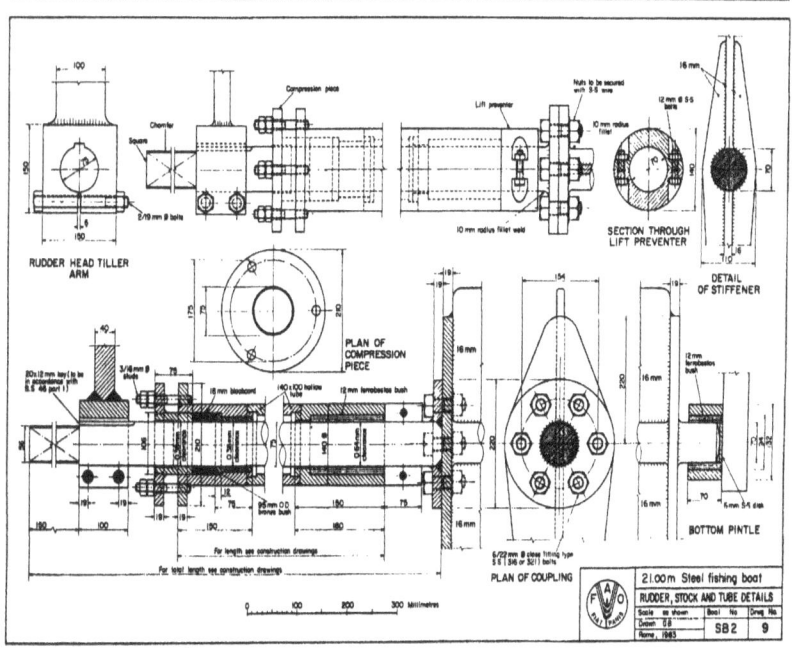

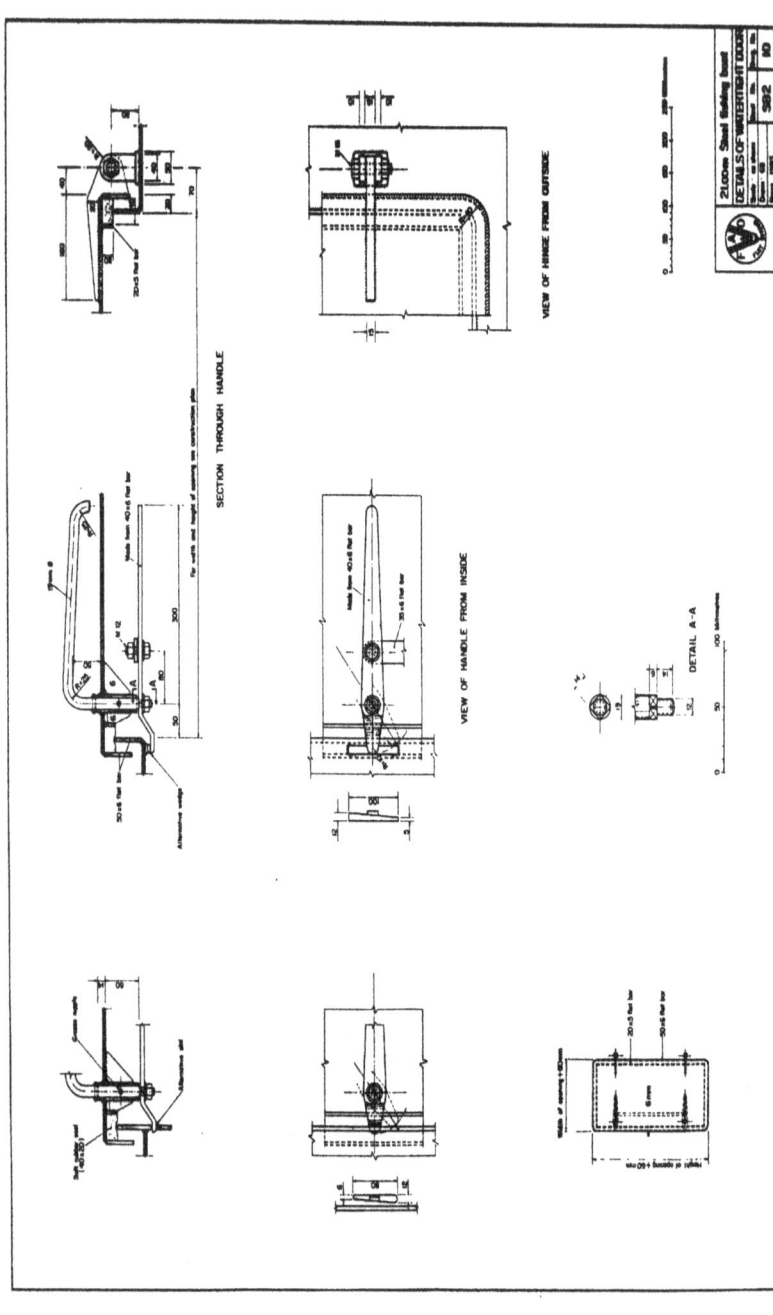

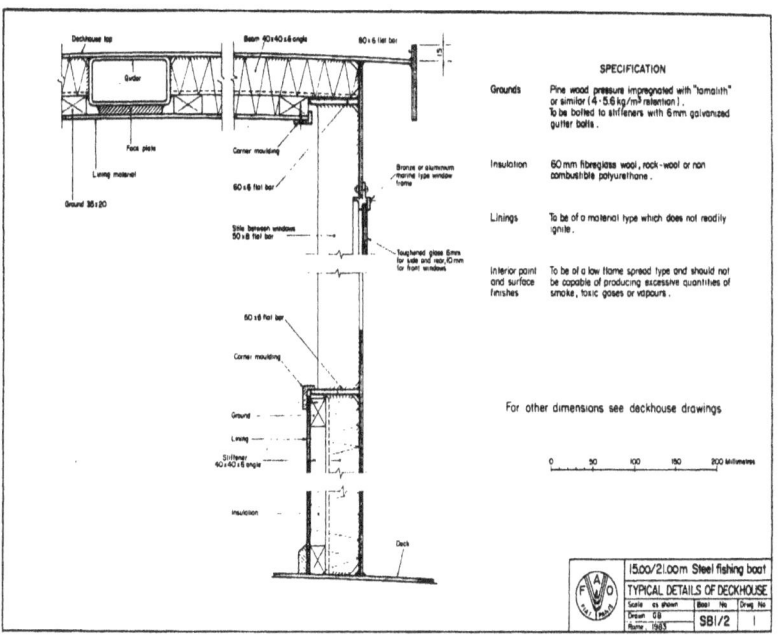

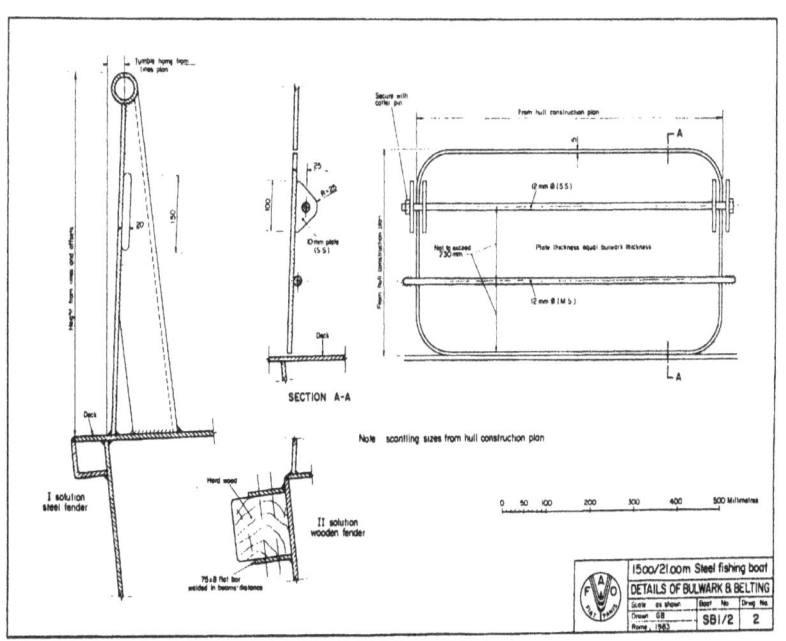

NO: 11308

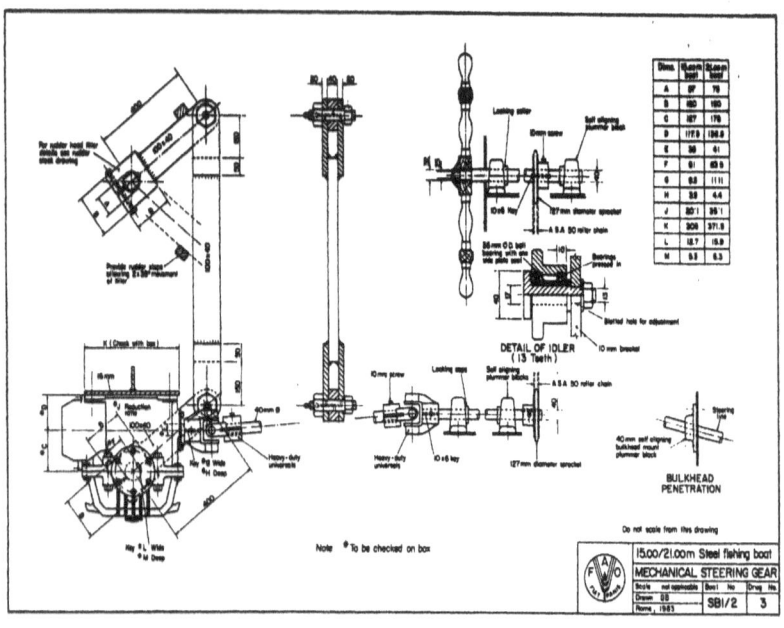

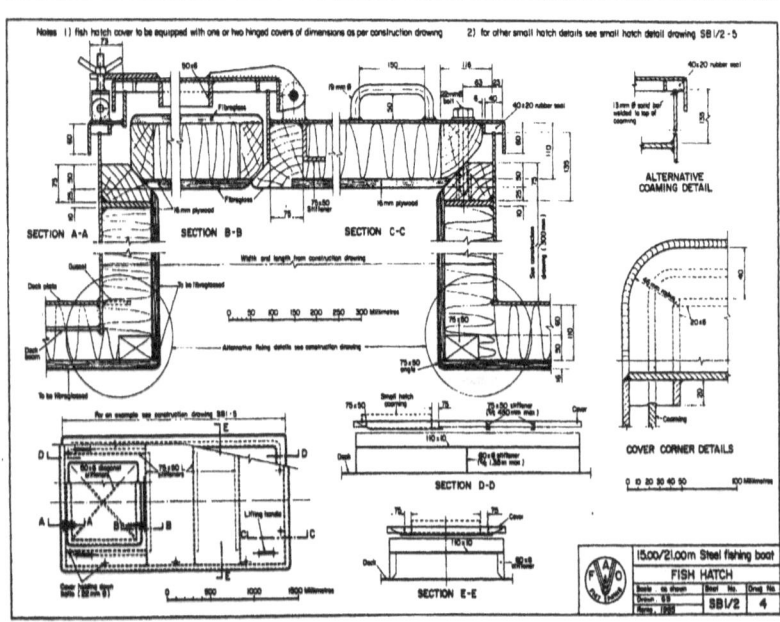

LOW COAMING
HIGH COAMING
END ELEVATION
HATCH PLANS
COVER CORNER DETAILS
Alternative hatch coaming fixing detail see construction drawing
Width and length from construction drawing
Width from construction drawing
40x20 rubber seal
40x20 watertight rubber seal
Deck plate
Deck beam
Gusset
Coaming
Hatch coaming
On deck fixing
Through deck fixing
Millimetres
Note cover plate thickness to be similar to deck plate (5mm min)
1500/21.00m Steel fishing boat
SMALL HATCH DETAILS
Scale as shown
Drawn GB
Rome, 1983
Boat No SBI/2
Drwg No 5

DETAILS OF WEB FRAME CONSTRUCTION
Deck beam gusset
Web frame
Chine gusset
Floor plate
Bottom
Normal penetration of longitudinals through web frames or beams
Notch in profile is also typical for notches in floor plates
Sealing plate (Plate thickness = bulkhead plating)
DETAIL OF PENETRATION OF LONGITUDINAL THROUGH O.T., W.T. BULKHEAD OR TANK END
Deck
Hull
Bulkhead stiffener
Bottom
DETAILS OF FREE ENDING LONGITUDINALS AND BULKHEAD STIFFENERS
Gusset thickness = longitudinal thickness
Stem bar
Fold (where necessary)
DETAIL OF LONGITUDINALS BRACKETED TO STEM BAR
Face bar fastened to longitudinal
Welded face bar
Bracket to be fitted to longitudinal
Pressed flange
A-A
B-B
END FASTENING OF INTERMEDIATE FLOORS FOR ALTERNATIVE FLANGING METHODS
1500/21.00m Steel fishing boat
DETAILS OF FRAMING SYSTEM
Scale not applicable
Drawn GR
Rome, 1983
Boat No SBI/2
Drwg No 6

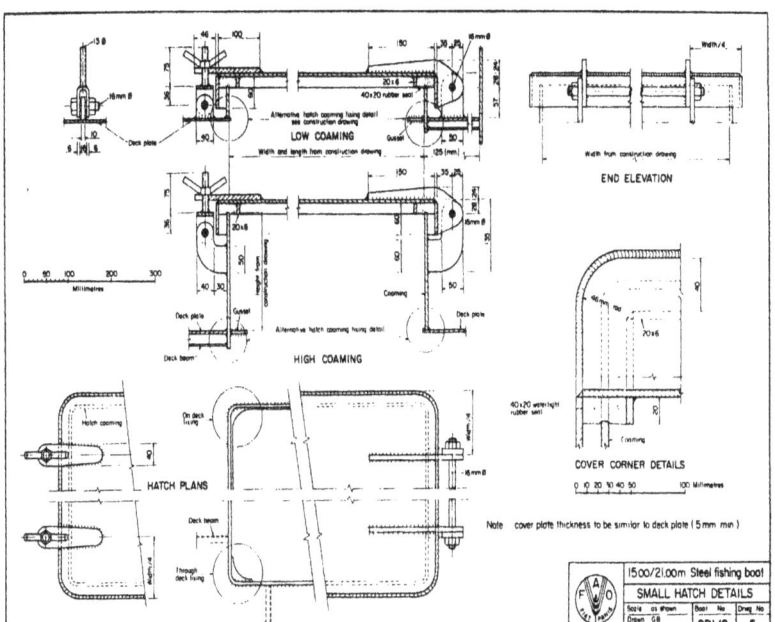

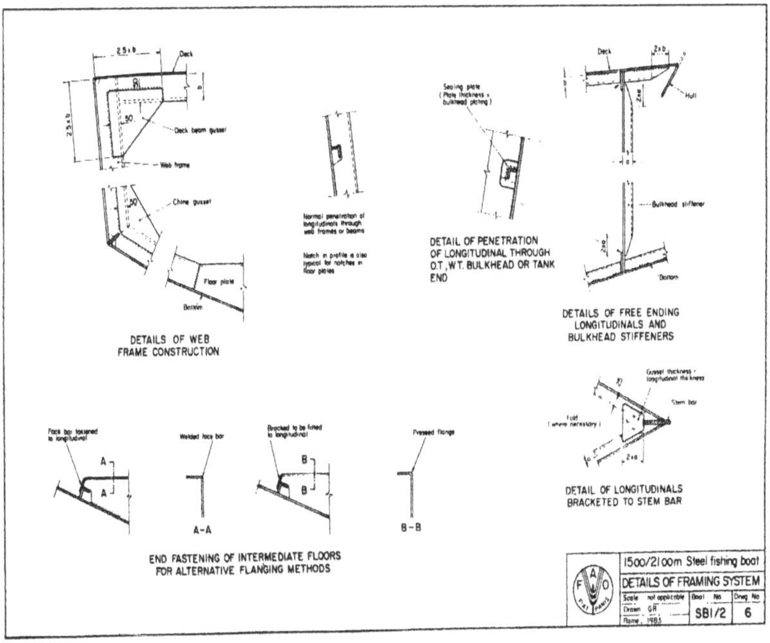

www.ingramcontent.com/pod-product-compliance
Lightning Source LLC
Chambersburg PA
CBHW020754230426
43665CB00009B/592